# Digging Our Own Graves

LABOR AND SOCIAL CHANGE,
a series edited by Paula Rayman and Carmen Sirianni

Barbara Ellen Smith

# Digging Our Own Graves

## Coal Miners and the Struggle over Black Lung Disease

TEMPLE UNIVERSITY PRESS PHILADELPHIA

Temple University Press, Philadelphia 19122.

Published 1987
Printed in the United States of America.

**Library of Congress Cataloging-in-Publication Data**

Smith, Barbara E.
Digging our own graves.

(Labor and social change)
Bibliography: p.
Includes index.
1. Lungs—Dust diseases—Social aspects. 2. Coal miners—Diseases and hygiene—United States. 3. Trade unions—Coal miners. I. Title. II. Series.
RC77.3.S55 1987 363.1'19622'3340973 86-14461
ISBN 0-87722-451-X (alk. paper)

The paper used in this publication meets the minimum requirements of American National Standard for Information Sciences—Permanence of Paper for Printed Library Materials, ANSI Z39.48-1984

Portions of certain chapters appeared earlier in article form: "Black Lung: The Social Production of Disease," *International Journal of Health Services* 11(3): 343-359, 1981, Baywood Publishing Co., Inc. (first reproduced under ths ame title by the International Institute for Comparative Social Research, Berlin, West Germany). "History and Politics of the Black Lung Movement," *Radical America* 17(2–3):89–109, 1983. "Too Sick to Work, Too Young to Die," *Southern Exposure* 12(3): 19–29. May-June, 1984. All are reprinted by permission. George Sizemore, "Drill Man Blues," in George Korson, *Coal Dust on the Fiddle: Songs and Stories of the Bituminous Industry* (Philadelphia: University of Pennsylvania Press, 1943), p. 450 is printed by permission of Mrs. Rae (George) Korson. Cover photo: Earl Dotter, American Labor Education Center.

To coal miners and their families all over the world

# Contents

# Preface

I first came to work for the Black Lung Association when I was barely 20 years old. It was the winter of 1971–1972. Extraordinary political changes were in the making: coal miners, miners' wives and widows were challenging powerful institutions that had once commanded their acquiescence—the hierarchy of the United Mine Workers, the coal operators' association, county political machines, the Social Security Administration. For a young college student from the Midwest, these developments in the mountains of West Virginia beckoned with a romantic excitement. Besides, the mountains were my ancestral homeplace; now I could return to them, not on a summer facation in the back seat of the family car, but on my own.

Working with the older coal miners and impatient young organizers who made up the Black Lung Association at that time was a formative political experience for me. Coming from a long line of subsistence farmers and preachers, I was instilled with a righteous, if vague, sense of populism that made me eager to work with "working people." But neither my political heritage nor my exposure to campus radicals had prepared me for what I found in West Virginia. It was the stark, ever-present reality and clear perception of class that struck me most. Virtually every coal miner over the age of 65 proudly claimed to have "fought in the battle of Blair Mountain with a machine gun" against the combined forces of coal company guards, the state police, state militia, and aerial bombers financed by the coal operators' association. I was dumbfounded.

Fortunately, it didn't occur to me to write about any of these experiences until my age and the changing times helped to deepen my understanding of what they meant. In 1978, more than six years after I had first worked for the Black Lung Association, I began the research for a dissertation on the black lung movement. The political atmosphere was altogether different. The reform administration of the United Mine Workers had disintegrated, the black lung movement had disappeared, and a storm of reaction was sweeping through the coalfields. The setbacks were frightening, but they made possible a more sober and critical perspective on the earlier period of upheaval.

I began this book as a labor history, asking obvious questions that seemed most important at the time: Why did the movement end this way? How did we fail? Who or what was to blame? As I dug deeper into the history of the black lung movement, however, these apparently clearcut questions began to seem ambiguous, even misleading. The assessment of whether the movement had succeeded or failed depended a great deal on whose goals were used as the standard of measurement—and goals varied considerably among different participants. Moreover, what the larger political culture defined as the movement's greatest victories and defeats often turned out to be largely symbolic events; they represented the visible outcomes of formal processes of reform (the passage of legislation, for example) but, in and of themselves, did not necessarily signify substantial and lasting change. The simplicity of my original questions faded as the labels of victory and defeat, success and failure, appeared more and more ephemeral. The central analytical problems increasingly seemed to involve the maddening complexity of social change itself, which had prevented any one person or group from controlling the course and outcomes of this movement.

As I delved further into the reforms sought and the controversies engendered by the black lung movement, it also became apparent that the movement was more than an important episode in labor history. At issue in the struggles over black lung was not simply the appropriate manner in which to resolve this social problem, but the very definition of this problem. Frequently, the most significant opponents that miners faced in these struggles were not coal operators or conservative politicians, but physicians. At the center of the black lung controversy was a profound ideological struggle between miners and physicians over the definition of disease.

As a result of these and other shifts in analytical emphasis, this book is a hybrid: part labor history, part political science, part medical sociology. It draws on diverse theoretical traditions in order to analyze not only the organization and development of the black lung movement, but also the history and conflict that underlay the movement: the social production of black lung in the workplace, the medical history of black lung, and the conflicting meanings which miners and physicians invested in the disease.

Today, as I write this preface, the political climate has changed once more. The "problem" of black lung has been popularly redefined to mean a government boondoggle, just another federal program that has wasted taxpayers' money through mismanagement, inappropriately liberal eligibility standards, and outright fraud. At the very least, this book may serve as a reminder that, not so very long ago, the problem of black lung was popularly defined very differently: as the needless suffering of thousands of coal miners, which constituted a national disgrace. Indeed, the history presented in the pages that follow suggests that the federal black lung compensation program became enormously expensive not because of waste or fraud, but because it revealed in a small way—in just one group of workers in one industry—the terrible extent to which human health is being destroyed in the workplaces of the United States.

# Acknowledgments

Written twice, first as a dissertation and later as a book, this text carries debts to many individuals. My academic advisors at Brandeis gave me freedom to follow my own sense of direction when I first set out to write a dissertation, but they also helped me get my bearings when I began to wander: Gordon Fellman, Charles Fisher, Jim Green, Ralph Miliband, George Ross, and Charlotte Weissberg. Many others took the time to read and criticize the manuscript in whole or in part: Grant Crandall, Penny Crandall, Michel Crétin, John David, Gail Falk, Leslie Falk, Gerd Göckenjan, Faith Holsaert, Lorin Kerr, Randy Lawrence, Pat Murray, Donald Rasmussen, Curtis Seltzer, Steve Shapiro, Meredeth Turshen, Greg Wagner, and Larry Zacharias. Several people allowed me to rummage through and borrow freely from their personal files and libraries: Gail Falk, Lorin Kerr, Tom Rodenbaugh, Curtis Seltzer, and Benita Whitman. Having access to their papers and books was enormously helpful in reconstructing the history of the black lung movement.

Support from various institutions gave me the luxury of free time to study and write. The Danforth Foundation supported my graduate work, and the U.S. Department of Labor, Employment and Training Administration, granted me funds to conduct research on black lung. The International Institute for Comparative Social Research gave me the opportunity to benefit from the intellectual stimulation of

European colleagues during a summer in Berlin, West Germany. More recently, the Southeast Women's Employment Coalition granted me a leave of absence to complete this book, and several staff members assumed an extra load of work to make that possible. I am grateful to all.

I owe a special thanks to Randy Lawrence, whose encouragement was steadfast; Meredeth Turshen, who willingly assumed the task of reading the entire manuscript at an early stage and pointing out analytical and stylistic deficiencies; Faith Holsaert, who took on the same task at a later stage; Lee Adler and Kristin Stevens, Cathy Schoen and Larry Zacharias, who gave me shelter and much more whenever I needed it; and Michael Ames, whose tolerance of my delays and enthusiasm for my efforts never faltered. Dan Peterson, my husband, earned several extra stars in his crown during my late and humorless nights at the typewriter; only my gratitude can match his patience.

Finally, I had many teachers in addition to those from academic settings. Some taught me about living with the disabling effects of black lung: Annie Bowen and Walter Bowen, who took me in as one of their own; and Ormond Perdue, who showed me hidden patches of ginseng and other secrets of Peeled Chestnut Mountain. Others who were active in the black lung movement welcomed me into their homes or workplaces, and recounted to me their histories. Their names appear below. I hope that this book does justice to the experiences and knowledge that all of these people so generously shared with me.

## Interviews*

Martin Amburgey
Willie Anderson
Bill Bailey
Rick Bank
Eric Brandt
Gina Brandt
Basil Brown
Don Bryant

Mike Burgess
Lyman Calhoun
Carl Clark
Eustace Clay
Clara Cody
E. E. Cody
Grant Crandall
Penny Crandall
Levi Daniel
Jesse Durham
Gail Falk
Walter Franklin
Robert Guerrant
Frank Hale
Leona Hall
Jim Haviland
George Hughes
Paul Kaufman
Lorin Kerr
Gibbs Kinderman
Joe Malay
Robert McDaniel
Warren McGraw
Arnold Miller
Woodrow Mullins
Joe Mulloy
Karen Mulloy
Milton Ogle
Willa Omechinski
John Pawlus
Robert Payne
Helen Powell
Frank Powers
Don Rasmussen
Tom Rhodenbaugh
Ernest Riddle
Craig Robinson
Norita St. Clair
Ray St. Clair
Sydnee Schwartz
Earl Stafford
Don Stillman
Bill Stone
Paul Stone
Lonnie Sturgill
Phoebe Tanner
Bill Weiss
Leslie Wellman
Teri West
Benita Whitman

* The interviews were conducted under terms of confidentiality, a decision that I later regretted. Unfortunately, I was unable to locate all respondents some six years after the interviews in order to request permission to reveal their names in association with specific quotes. Therefore, all excerpts from the interviews remain anonymous.

# Digging Our Own Graves

# 1

# The Medical Construction of Black Lung

In April of 1831, Dr. James C. Gregory of Edinburgh, Scotland, performed an autopsy on a recently deceased patient. John Hogg, the deceased, had been a soldier most of his adult life, but for the last ten to twelve years he had worked in the coal mines at Dalkeith. Prior to his death, Mr. Hogg had come to Dr. Gregory with complaints of breathlessness and chest pain, which had forced him to quit his occupation as a miner. His symptoms soon worsened to include a severe cough, which at times yielded a dark and viscous expectoration. John Hogg was admitted into the infirmary at Edinburgh on March 29, 1831. Twenty days later, he died.

When Dr. Gregory cut open the dead man's lungs, he saw the following:

> Both lungs presented one uniform black carbonaceous colour, pervading every part of their substance. The right lung was much disorganized, and exhibited in its upper and middle lobes, several large irregular cavities, communicating with one another and traversed by numerous bands of pulmonary substance and vessels. These cavities contained a good deal of fluid, which, as well as the walls of the cavities, partook of the same black colour.[1]

Although he acknowledged that such changes in lung color could be associated with dark pigment disorders ("melanosis"), Dr. Gregory speculated that John Hogg's black lungs were a result of his occupation. More specifically, Gregory argued that the condition derived from "the habitual inhalation of a quantity of the coal-dust with which the atmosphere of a coal-mine must be constantly charged."[2] It is worth noting that John Hogg attributed his respiratory troubles to working in the mines, and that Dr. Gregory found this information sufficiently credible to include it as his final argument in support of occupational causation.

Dr. Gregory's "Case of peculiar Black Infiltration of the whole Lungs, resembling Melanosis," was the first of several reports in British medical journals that tied the respiratory problems and blackened lungs of coal miners to their occupation.[3] A few decades later, when underground coal mining began in earnest in the United States, similar accounts began to appear in this country's medical literature. In 1869, Dr. John T. Carpenter of Pottsville, Pennsylvania, located in the heart of the anthracite district, reported to his county medical society concerning the respiratory disease and other occupational health problems of coal miners. He, too, remarked upon the black sputum, chronic cough, and "penetration of solid, carbonaceous matter into the lungs [which] we all have verified by post-mortem examinations."[4] Carpenter concluded that the daily inhalation of dusts and gases in the workplace was "the most serious" source of occupational disease in the industry. Although he offered no suggestions for reducing these health risks to coal miners, Dr. Carpenter did have a few words of caution for insurance companies: "I could not conscientiously advise any life insurance company to do business among miners, except on short periods of risk, and at large increase of percentage."[5]

This mid-nineteenth-century discussion of black lung disease was not confined to the obscure literature of the medical profession. Coal miners in Britain and the United States clearly believed in the occupational origin of certain respiratory problems, as evidenced in physicians' records and in reports of underground working conditions.[6] Popular novels and public records of the time also suggest that, at least among those familiar with coal mining regions, recognition of black lung disease was widespread. Friedrich Engels, for example, utilized the *First Report* (1842) of the Children's Employment Commission to construct a detailed description of coal miners' occupational health, including "black spittle" disease, in his treatise *The Condition of the Working Class in England*.[7] Emile Zola's great novel of the coal industry, *Germinal,* depicts a victim of black lung in the character Bonnemort.[8] These and other social critics invoked coal miners' underground labor, lung disease, and premature deaths as powerful symbols of the dark and dangerous human effects of a rising industrial capitalism.

And yet, by the turn of the century, the medical profession grew silent on the subject of miners' occupational lung disease. In Great Britain, one hundred years after Dr. Gregory "discovered" a coal miner's black lungs, physicians did not generally recognize or accept as legitimate *any* disabling occupational lung disease associated with the inhalation of coal dust. In the United States, one hundred years after Dr. Carpenter concluded that dust and gas inhalation was the most serious occupational health problem in the Pennsylvania coal industry, physicians rallied behind the position that "the control of coal dust is not the answer to the disabling respiratory diseases of our coal miners."[9] In both countries, recognition of black lung as a legitimate, disabling, and compensable occupational disease required the active intervention of coal miners themselves.

The medical history of black lung does not coincide neatly with conventional notions of scientific discovery as a "march of progress" toward truth. Over time, physicians have constructed three distinct definitions of miners' occupational lung disease which are, to some extent, mutually contradictory. They cannot be explained by any logic internal to science or by any change in the nature or prevalence of disease.[10] Physicians' shifting perceptions of black lung ultimately find their most satisfactory explanation in medicine's historically changing view of the human body and disease; this in turn has been contingent upon social factors ordinarily considered extrinsic to scientific thought. Thus, the medical history of black lung entails the social history of medicine.

## Dissect the Body, Determine the Disease

When Dr. James C. Gregory developed his arguments in 1831 for an occupationally caused "spurious melanosis," or black lung, he was working within a medical cosmology that was only a few decades old. The birth of modern medicine is typically dated from the last years of the eighteenth century and is placed in western Europe, specifically France. It was here that physicians effected a fundamental transformation in their perception of the human body: whereas the mid-eighteenth-century view was of a unitary whole, a single enclosure of living organs and tissue, fifty years later physicians broke down that awesome vitality and objectified the body as a system of functionally interrelated parts. No longer were physicians bound to a view of disease as a complex of related symptoms; now they sought the internal workings of disease through the identification of localized pathologies and the correlation of external symptoms with morbid events in specific tissues.[11]

As a historical event, the breaking down of the human body was an act of domination, both perceptual and social. It first took place within the Paris hospital system, which was designed to serve the indigent population of the city, and was reorganized under state control in the course and name of the Revolution.

Centralization and expansion of the urban hospitals under a single authority (as opposed to diverse private charities, more typical of the time) made it possible for physicians to examine and classify great numbers of "cases" and to apply statistical techniques in their clinical research. More fundamentally, however, these political reforms gave physicians access to the bodies of individuals who, by virtue of their class position, were relatively powerless in the doctor-patient transaction. Unlike eighteenth-century medical practice, which was typified by a patronage relationship between wealthy aristocrats and their socially inferior physicians, in the Paris hospitals, physicians clearly had the upper hand. They were able to expose, examine, open up, dissect, and ultimately redefine the human body in the context of a class relationship with the poor.[12]

This was the golden age of pathological anatomy. "Open up a few corpses," declared the French physician Bichat, and "you will dissipate at once the darkness."[13] Although it did not produce many therapeutic innovations, the early nineteenth century was a transitional period of exploration and speculation. Restricted in diagnostic methods for the living patient, physicians like Dr. Gregory combined the perceptual breakthroughs and post-mortem investigations of the new anatomy with more conventional attempts to elicit and classify the patient's symptoms. The result was a medical construction of black lung that coincided with miners' own perception of the disease.

Across the Atlantic, in the United States, eclecticism and diversity characterized medical training, knowledge, and practice throughout the 1800s. Indeed, professional consensus eluded physicians well into the twentieth century. Although a select group obtained education in medical schools and apprenticeships in hospitals of the United States and Europe, many physicians had little formal training and faced few legal restrictions on their practice. Especially in rural areas, where most of the population, of course, lived, physicians practiced "shopkeeper" medicine, dispensing therapies, advice, and comfort to the people of their small town and the surrounding countryside. The uncertain financial and social status of these private practitioners lent a relative equality and negotiability to the doctor-patient relationship, though the regimentation and subordination of patients, especially charity patients, were features of the emerging urban hospitals.[14]

Within this diverse medical climate, physicians in the small towns of the eastern coalfields first identified a disabling lung disease among coal miners and attributed it to their underground occupation. The medical records available from the time suggest that physicians' understanding of black lung varied. Although they all conducted pathological investigations and emphasized these findings, some reasoned from the standpoint of the clinician: they observed miners' persistent cough and black sputum and speculated that they resulted from the irritating effects of dusts and gases inhaled in the workplace; "miner's asthma is chronic bronchitis," they argued.[15] Coal miners in their care often lived many years with these symptoms, but their chronic condition predisposed them to other respiratory ailments, and many died from acute pneumonia. Other physicians stood on pathological evidence: they argued that miners' respiratory problems derived from the inhalation and deposition of carbon in the lungs; this mechanical accumulation led to "carbonaceous solidification in the air cells," as one physician concluded from post-mortem examinations. "The Scotch people call it spurious melanosis, really a coal miners' consumption. I have no doubt the carbonaceous particles caused their death."[16]

These initial perceptions of black lung contrast vividly with subsequent medical definitions of the disease. The early accounts took as their starting point the perceptible condition of the patient; his reports of breathlessness and the observable signs of cough and sputum were granted legitimacy as evi-

disease. Diagnostic technology was primitive; the physician relied largely on what the unassisted eye could see, and the ear, extended by the stethoscope, could hear. The medical cosmology in which physicians reasoned and interpreted was eclectic, speculative, open-ended. There was no rigid division of labor between the theory and the practice of medicine, as the clinician was also a researcher. The biological world of the body and the social world of the patient were not divorced; the search by physicians for the causes of disease extended into the environment of life and work. (Thus, Dr. John Carpenter's report to his county medical society included a frank and straightforward account of the many hazards of underground coal mining.) Finally, the structure of medical practice as a small, private business, and its uncertain status in a period when lay healing persisted and the ideological authority of science was not fully established, rendered physicians at least minimally accountable to the opinions of their patients.

All of this soon would change. During the last half of the nineteenth century, the application of the natural sciences in the laboratory study of the human body yielded new understandings of human physiology, biology, and histology. Building on the precepts of cell theory and the discovery of bacteria, laboratory researchers in Germany focused their attention at the level of the cell and sought to correlate specific aberrations in function with discrete agents of disease. The resulting germ theory held that every disease is caused by a specific bacterium or agent. This theory was soon utilized as a fruitful scientific tool for elucidating at least those diseases that were compatible with its implicit assumptions; it was also wielded as an ideological weapon against those who emphasized the social origins and prevention of disease. Germ theory asserted that the essential process of disease unfolded entirely within the individual—indeed, within individual cells. By confounding the microscopic *agent* with the *cause* of disease, this theory offered a wholly biological perspective on the nature of illness.[17]

The triumph of scientific medicine was associated with fundamental changes in the production of medical knowledge. Research gradually became the province of biomedical scientists, who, because they were located in the laboratory, were separated from the human patient. Diagnosis became a process of identifying distinct disease entities, and confirmation of the diagnosis was sought in quantifiable medical procedures and tests. Patients' own testimony concerning their condition was relegated to a distinctly secondary, even suspect, status. Scientific medicine in its laboratory phase involved what one author termed the "disappearance of the sick-man" from the medical view.[18] The patient came to appear almost incidentally as the medium for disease, eclipsed by the focus on identifying discrete pathologies. Absent an identifiable clinical entity, or at least a quantifiable deviation from physiological norms, the patient was by definition pronounced healthy. In that case, protestations of feeling ill became a matter for the psychiatrist.

At the same time medical knowledge was being reconstructed on a biological basis, many of the institutional features that now characterize the medical profession began to emerge.[19] A spirit of unity gradually overcame the bitter rivalries among divergent schools of medical thought (homeopathic, allopathic, herbalist), as educated physicians discovered a common enemy in the untrained practitioner. Alliances of physicians within the different states succeeded in erecting legal barriers to professional entry via licensing laws and the formation of state medical boards, which they controlled. At the beginning of the twentieth century, the American Medical Association (AMA), which had been established in 1846, began to consolidate its organizational power around the related issue of medical education. Licensure was insufficient to shore up professional status, standards, and earnings if large numbers of people, especially lower-class people, were able to attend medical school. Already threatened by the increasingly powerful state medical boards, which could refuse to recognize the diplomas of their graduates, many small medical schools were forced to close in the wake of the famous Flexner report on medical education, released in 1910. Initiated by the AMA and funded by the Carnegie Foundation, the report deplored the absence of a rigorous scientific base in medical education and accordingly prescribed specific changes in curricula. The conceptual framework and institutional mechanisms were now in place to enforce professional conformity in medical theory and practice. This union of institutional power and ideological authority soon made medicine the most elite and lucrative professional domain in the United States.

The turn of the century also saw important changes in the dynamics of class relations in the coal industry which directly affected medical practice in mining towns. The locus of production and economic power in the bituminous industry began to shift from the traditional Central Competitive Field (Ohio, Indiana, Illinois, and western Pennsylvania) to the newly opened, nonunion mines of central Appalachia. In the rural, sparsely populated mountains of southern West Virginia, eastern Kentucky, and areas farther south, investors erected towns from the ground up and extended the power that ownership confers from the workplace throughout the community. These were company towns, where nothing—not churches, police, or medical care—escaped the scope of corporate domination. Here, physicians elaborated a completely different concept of black lung, one compatible with both the new framework of scientific medicine and the economic imperatives of corporate control. For the next fifty years, the occupational disease and disability associated with the inhalation of coal dust were absent from the literature and concerns of medicine.

## The Bitter Legacy of Silence

The towering forests and vast underground coal seams of central Appalachia remained largely undisturbed during the first century after the American Revolution. Scattered families of subsistence farmers populated the rugged mountains, but their demands upon the area's natural resources tended to be modest, based on local need rather than commercial exchange. The last three decades of the nineteenth century, however, were a time of tremendous economic growth, when the domain of industrial capitalism extended its boundaries across the entire nation. Demand for coal soared, and the mineral wealth of central Appalachia lured investors to exploit the entire region.

The tale of this initial exploitation has been told vividly and thoroughly elsewhere.[20] Railroad companies were the first to penetrate the fortress of mountains during the 1870s, but in their tracks came a series of land speculators, prospectors, and businessmen. In the space of twenty years, over half the mineral wealth in many counties became concentrated in the hands of a few nonresident individuals and companies.[21] Lacking a sufficient labor force in the sparsely populated mountains, these investors dispatched recruiters, who advertised in Europe, traveled to urban centers of the North, and even emptied southern jails in their efforts to locate male owrkers. Those who had tilled the soil came to mine the earth—ex-slaves from the South, native mountaineers, former peasants from southern Europe—and they formed a polyglot mixture in the new coal camps.[22]

The absence of towns and infrastructure required companies to construct entire camps to house and service these workers and their families. They turned this practical necessity to economic gain, however, through ownership and commercial operation of stores, housing, and the health care system. Miners were typically paid in nonlegal tender, or scrip, which could only be traded for its full value at the company store. The profits to be made from these secondary commercial ventures could be quite significant for a company's economic survival in the highly competitive coal industry. Control of housing, police, and railroad access to the towns also enabled the operators to forestall what they perceived as a pernicious threat to their economic status: unionization.

Although there were exceptions, the living conditions in these company towns were often miserable. For between six and eleven dollars per month, the more unfortunate received a decrepit house in a bleak setting:

> In the worst of the company-controlled communities the state of disrepair at times runs beyond the power of verbal description or even photographic illustration, since neither words nor pictures can portray the atmosphere of abandoned dejection or reproduce the smells. Old, unpainted board and batten houses—batten going or gone and boards fast following, roofs broken, porches staggering, steps sagging, a riot of rubbish, and a medley of odors—such are the worst camps.[23]

Sanitation facilities often reflected a gross indifference to public health. Houses were erected one next to the other, with outhouses situated in close proximity to the back doors and, in some cases, to the water supply. When the militia entered Paint Creek, West Virginia, in 1912 to quell the class warfare that had erupted there, sanitation was so poor that one of the troops' first assignments was to spread lime through the camps.[24] Subsequent U.S. Senate hearings concerning the living conditions at Paint Creek uncovered evidence of serious public health problems associated with the layout of coal camp housing, sanitary facilities, and the water supply. Typhoid fever and gastroenteritis were common.[25]

Ten years later, a survey of health conditions in the bituminous coalfields conducted by the U.S. Public Health Service confirmed that the situation at Paint Creek was the rule and not the exception:

> The fact that manure is a fly-breeding material of first importance is practically unrecognized in the places surveyed. Ordinances or requirements for the systematic and frequent removal of manure are conspicuously absent. . . . Screening against flies may be said to be generally inadequate. . . . Control of disease carriers and communicable disease contacts appears to be an unknown art.[26]

The community, of course, was not the only locus of disease; underground coal mines were industrial battlefields where injury and death were daily events (see chapter 2). Between 1906 and 1935, over 45,000 miners died in "accidents" and explosions in the underground coal mines of the United States. During World War I, the death rate of miners in southern West Virginia exceeded that of the American Expeditionary Force.[27] According to one conservative estimate, the rate of serious injury may have been eight times higher than the death rate.[28] And even that is an underestimate of the carnage, for it does not include the toll from occupational disease. There are no reliable, industrywide statistics on occupational injuries or illnesses for this early period, but hospital records are suggestive. A random sample of miners admitted in 1902 to the hospital in Fairmont, West Virginia, received the following diagnoses: fracture and dislocation of the spinal column; compound fracture of the ribs, penetrating lung; compound dislocation of the leg; contused wound of the head and face; contused wound of the arm; and compound dislocation of the thigh.[29]

Even as they shaped the production of death and disease, the coal companies sought to control the definition and treatment of medical problems. The company doctor system that the coal industry instituted resembled what is now termed a prepaid health plan, but with a repressive twist: the physician was hired by and accountable to the company, yet miners were required as a condition of employment to pay for his services. Some companies even turned a small profit from this arrangement by pocketing 5 to 10 percent of the medical checkoff from miners' pay.[30]

The most comprehensive study of the coalfield company doctor system was undertaken in 1946 at the behest of the United Mine Workers by the Coal Mines Administration of the U.S. Department of the Interior. Their findings were published as the Boone report, named for its author, Rear Admiral Joel T. Boone of the U.S. Navy.[31] Boone and his five teams of field researchers found that in 1946, 97 percent of the miners in southern West Virginia, eastern Kentucky, Virginia, Alabama, and Tennessee were covered by a "prepayment" plan for medical services. Virtually all of these miners also prepaid for hospitalization.[32] In only 6.5 percent of the 260 mines surveyed were there "excellent dispensaries, which [were] utilized for industrial medicine and ordinary medical care of miners and their dependents."[33] More often, the company doctor's office was drab and poorly equipped:

> Some of these offices are neat and well-equipped, but at least half are unattractive, meagerly furnished, and fitted with scarcely any more equipment than a general practitioner's bag contains. In 13 instances it was noted that the doctor's office was "very disorderly" or "dirty." At three of the mines visited, the doctors' offices were described as "insanitary."[34]

The implications of the company doctor system ranged beyond the quality of health care, for only physicians whose medical philosophy did not challenge the economic interests of the companies remained in practice. Above all, this meant that company doctors ignored or redefined as individual faults the manifest economic origins of disease. For example, physicians who testified at the Paint Creek hearings downplayed the threat from typhoid fever and blamed miners and their families for the failure to maintain company housing:

> SEN. MARTINE: You wouldn't consider it sanitary in your own home to have an open vault with festering putrid matter within 40 feet of your kitchen door, and flies crawling over your food.
>
> DR. ASHBY: The people are advised to get screens.
>
> SEN. MARTINE: There are many instances when that does not exist?
>
> DR. ASHBY: They do not purchase them, probably.[35]

Company physicians also tended to ignore or to ascribe to the individual the manifold occupational hazards in the company-owned mines. Granting medical attention to occupational danger and disability could have seriously undesirable economic ramifications for the coal companies: it could lead to demands for preventive action in the workplace and to coverage of a new disease under the workers' compensation system, thereby increasing corporate liability. Company doctors typically were called upon to testify in the operators' favor whenever miners filed workers' compensation claims. For physicians to increase their employers' liability by investigating the occupational etiology of disease would have run completely counter to the economic logic of this system. Instead of acknowledging occupational or economic causation, the company physicians' implicit philosophy on the origins of disease was individualistic and behavioristic. Industrial accidents were attributed to individual "carelessness," and illness, to self-destructive personal habits like alcoholism.

In the context of this medical cosmology and social environment, physicians developed a new understanding of black lung. They did not simply avert their gaze from miners' respiratory distress, as many have assumed. Rather, physicians dubbed the widespread breathlessness, expectoration of sputum, and prolonged coughing fits "miners' asthma," a benign condition that was to be expected. They thereby constituted the symptoms of lung disease as a *norm*—for coal miners. As one physician in Pennsylvania commented in 1935: "As far as most of the men in this region are concerned, so called 'miners' asthma' is considered an *ordinary* condition that needs cause no worry and therefore the profession has not troubled itself about its finer pathological and associated clinical manifestations" (emphasis added).[36]

Recent exposés of corporate efforts to cover up evidence of occupational disease have often revealed a conspiracy of silence between company physicians and their corporate employers.[37] The situation in the coal industry with regard to black lung was far more subtle, however. There is no evidence of deliberate conspiracy; rather, physicians obliterated the disease in the very content of its definition. This simultaneous recognition and effective denial of black lung had insidious and far-reaching implications. If disease was "natural" and inconsequential, then prevention was unnecessary. The new definition also enabled physicians to blame disease on the individual miner. Those who professed disability from this "benign" ("needs cause no worry") condition were viewed with suspicion, and their impairment was attributed to malingering or to an irrational "fear of the mines." The suffering and disability that attended black lung, thus, became medically stigmatized signs of psychological duplicity or instability.

The economic liaison between coal companies and physicians offers the most obvious and persuasive explanation for the reconstitution of black lung as a benign miners' asthma. But a probe into the coalfields and beyond reveals many other factors. The company doctors' evaluation and treatment of mining families' health was not simply a medical transaction; it was a class relationship. The rural monoeconomy of the Appalachian coalfields generated a rather simple and vivid class structure in which physicians, mine superintendents, and a few other professionals and businessmen formed a small, elite island in a working-class sea. The division between these two classes was everywhere apparent, embodied in physical geography and human characteristics: the elite's spacious white houses on the hill were set apart from the miners' small dwellings along the creekbank; the clean physician dressed in jacket and tie differed markedly from the dusty and begrimed miner after a day's work; the educated language and cultural refinements of the wealthy were distinct from the slang, foreign tongues, and illiteracy of the working class. The degraded social environment of the company towns not only was endured by but also was attributed to coal miners and their families. That miners did not put screens on their rented houses only demonstrated their ignorance and depravity; that many had a chronic respiratory condition only confirmed their difference. What was "normal" for miners and their families was by no means normal for the company doctor.[38]

Extending the analysis beyond the southern coalfields suggests parallels between company physicians and other, independent, medical practitioners. Most importantly, they held in common the theoretical framework of scientific medicine. Imbedded in the very content of this scientific world view were certain concepts that also mitigated against the recognition of black lung: the individualistic, biological perspective on illness, with its attendant tendencies to blame the victim; the neglect of social and economic factors in disease causation, and associated ignorance of occupational conditions and hazards; the emergent emphasis on "objective" diagnostic tools, and the related denigration of patients' own perceptions of their illnesses. Viewed within this broader context, the financial ties between physicians and coal companies appear insufficient as a thorough explanation for the new definition of black lung; the social and medical framework that company doctors shared with other physicians also informed their perception and construction of disease.

The historical record supports this assertion. The few relevant documents written by individual physicians between 1890 and 1940 reveal striking similarities in their perceptions of coal miners' health and illness, regardless of geographic location or medical practice arrangements. For example, in 1919 one physician, Dr. Emery Hayhurst from Ohio, who also held a Ph.D. and was a university professor and a consultant in industrial hygiene to the state Department of Health, authored a report on "The Health Hazards and Mortality Statistics of Soft Coal Mining in Illinois and Ohio." In spite of his own detailed explication of the numerous dangers in the coal industry, which included dust, gob piles, electricity, poisonous gases, and contaminated water supplies, he managed to conclude: "Housing conditions, and hurtful forms of recreation, especially alcoholism, undoubtedly cause the major amount of sickness. The mine itself is not an unhealthful place to work."[39] Dr. Hayhurst documented that "violence" was by far the most important cause of death for Illinois coal miners, taking their lives four times as often as it did the lives of other adult males in the area. Nevertheless, he did not include in his eleven recommendations preventive action in the workplace. He did, however, suggest "promotion of Americanization, especially in districts of foreigners," as well as medical licensure, "as a check against unscientific methods."[40]

The few clinical studies of miners' lung disease during this period were also carried out within the existing conceptual framework of scientific medicine—with negative consequences for the recognition of black lung disease. From 1890 through the end of World War II, there were five reported clinical surveys of lung disease among coal miners, three of which were conducted by the U.S. Public Health Service.[41] The straightforward (if crude) observations of nineteenth-century physicians, who reasoned from evidence of occupational dust inhalation, clinical symptoms of respiratory distress, and pathological findings of black lungs, differed from the observations of these twentieth-century researchers, who operated within a far more narrow and stringent framework. They sought a specific etiologic agent, exposure to which would yield physical impairment and evidence of which could be confirmed by objective diagnostic techniques. Their research was also informed by the conviction that silica dust was by far the most pernicious in its effects on the respiratory system; the fibrotic changes characteristic of silicosis were taken as the standard for respiratory impairment from dust inhalation.

All of these studies relied heavily on the X-ray as a diagnostic tool. This was supplemented with an occupational history, in most cases with clinical exams, and in one case with medical records of previous respiratory illness. Unlike their nineteenth-century predecessors, none of these researchers attempted to correlate pathological findings upon autopsy with clinical evidence during life. The studies varied in quality, number of miners included, and so on, but they all reached the same conclusion: coal miners exposed to the specific agent of silica were at risk of an occupational respiratory disease, anthracosilicosis, but those exposed only to "ordinary" coal mine dust were not. Several authors invoked a few pathological and experimental studies to bolster this conclusion: "Observations upon the lungs of bituminous coal miners and of experimental animals suggest that, when pure coal dust is inhaled, even in large quantities, it produces little or no fibrosis."[42]

On one level, these physicians failed to discover a "new" disease because they were not looking for one. Convinced by the mechanical perception that hard and sharp-edged particles of silica must be more deleterious than the soft and rounded carbon, they tended not to seek an occupational disease associated with coal dust alone.[43] A few explored this possibility, but their standard of respiratory harm was based on the action of silica. At a deeper level, these studies did not uncover black lung because the disease was not readily accessible within the paradigm of germ theory and the requirements of objective (i.e., X-ray) evidence. It is known today that respiratory disability in coal miners correlates poorly with X-ray evidence of disease, and that X-ray evidence correlates imperfectly with pathological findings.[44] There is no simple and direct relationship between exposure to a single agent, clinical symptoms of disability, physical tests showing lung function impairment, and X-ray evidence of disease. In other words, black lung remained invisible, even to those who deliberately studied respiratory disease among coal miners, because the scientific framework in which they operated mitigated against its recognition.

So it happened that, a century after Dr. James C. Gregory recorded his conclusions concerning the occupational origins of John Hogg's respiratory trouble, the medical profession in the United States systematically denied the existence of a distinctive, seriously disabling black lung disease among coal miners. The tangled strands of social and medical factors that were woven together in the trivialization of black lung were not unraveled by physicians alone, or by any developments internal to science. It was coal miners themselves who acted to redefine their own health and disease. During two long and bloody decades, miners and their families in the southern coalfields, especially West Virginia, collectively struggled against the "oppression of the coal barons," as they put it.[45] Their eventual victory in unionization produced a new structure of industrial relations in the coalfields, a new voice for miners in matters of occupational safety and health, a new breed of medical practitioners in mining towns, and a new definition of black lung disease.

## "Now You See It, Now You Don't"

After two decades of bitter armed conflicts in their company towns, the southern bituminous coal operators finally capitulated in 1933 to unionization and collective bargaining with coal miners. The political atmosphere generated by the Great Depression and the New Deal lent support and legitimacy to miners' and other workers' organizing efforts; legal barriers and ideological objections to unionization began to crumble. By 1934, the United Mine Workers of America (UMWA) was 400,000 strong, and its new base of power lay primarily among the militant rank and file of central and southern Appalachia.[46]

Unionization and collective bargaining did not pacify labor-management relations or bring peace to the coalfields, however. The period during World War II and immediately after was one of the most turbulent in the long history of coal mining in the United States. Refusing to acquiesce to the wartime no-strike pledge, UMWA president John L. Lewis and nearly half a million coal miners repeatedly shut down the entire industry. In response, the federal government seized control of the coal mines on four separate occasions. In 1943 alone, more than seven million worker-days of production were lost to walkouts, and twice President Roosevelt federalized the operators' coal properties.[47]

In 1945, Lewis brought to collective bargaining sessions a controversial demand that miners had supported through petitions and convention resolutions for at least fifty years. At issue was the company doctor system. Miners wanted a health care plan organized on a completely opposite basis from that of corporate-controlled medicine—a plan that was industry financed and union controlled. Lewis proposed that coal companies contribute ten cents per ton to the union "to provide for its members modern medical and surgical service, hospitalization, insurance, rehabilitation and economic protection."[48] Failing to win this, the union bargaining team returned to the annual contract negotiations in 1946 with industry-financed health care as its number one demand. The operators again refused, and miners struck en masse when their old contract expired. Once again, the government federalized the coal industry.

Lewis proceeded to take full advantage of his new adversary at the bargaining table, Interior Secretary Julius A. Krug, and was finally able to wring agreement to a union health care plan, which was to be financed by a royalty on mined coal of five cents per ton.[49] This concession marked only the beginning of the controversy, however. The operators continued to resist implementation of the fund, and they viewed the royalty payments as an unjust ransom required to regain their federalized property.[50] During the next four years, UMWA miners struck repeatedly, despite court orders and fines totaling close to five million dollars, in order to force coal companies to comply with their contractual agreement to establish the fund.[51]

The 1950 contract negotiations pitted John L. Lewis against George Love, who presided over the largest coal company in the United States, Pittsburgh-Consolidation Coal.[52] The impact of their landmark agreement reverberated through the coalfields for decades, as it transformed relations between UMWA leaders and the big operators, drastically altered the work process through mechanization, and slashed the size of the work force (see chapter 2)). In exchange for the unimpeded introduction of machinery into the mines, the larger operators pledged their commitment to an industry-financed and union-controlled Welfare and Retirement Fund, designed to provide medical care, hospitalization, and pensions for miners and their families. The fund was administered by a tripartite board, composed of one member chosen by the operators, one selected by the union, and a third, neutral, member chosen by the first two. With Lewis as the UMWA representative and Josephine Roche, a former coal operator and close Lewis confidante, as the neutral trustee, the union was easily in a position to shape the benefit program that the fund was to build. Union control was further consolidated by the appointment of Roche as fund administrator.[53]

After 1950, opposition to the fund originated less from coal companies and their organizations and more from physicians and the American Medical Association. Although the fund's medical staff initially upheld such sanctified traditions as fee for service, the reaction from the medical establishment was strident. Physicians attacked the fund as a subversive center of socialized medicine; they feared encroachment on their exclusive discretion over the delivery of health care, and they charged that miners "seemed to want to run the whole show."[54] The irony in this attack on the fund was not lost on those familiar with the history of coalfield health care. Physicians who had never questioned the company doctor system now did battle with the fund, professing concern that it denied patients a "free choice of physician."

Despite organized opposition from county and state medical societies and the American Medical Association, whose tactics included expelling cooperating physicians and denying them hospital privileges, the fund was able to build a comprehensive, cradle-to-grave health care system for coal miners and their families.[55] Although primarily a financing mechanism rather than a direct service provider, the fund represented an enormous financial lever with which to influence the quality and availability of health care in the coalfields. In 1952, for example, the fund established the Miners' Memorial Hospital Association to construct, equip, and direct ten hospitals in Kentucky, West Virginia, and Virginia. When beneficiaries required special medical attention that was unavailable in the region, as sometimes did miners disabled by industrial accidents, the fund paid for transportation and the services of specialists.

Other avenues of fund influence were less tangible but nonetheless important. Physicians and public health advocates who had cut their political teeth in the left-wing atmosphere of the 1930s and subsequently staked many hopes on the unsuccessful effort to establish a national health system found in the fund a new beginning, a source of renewed optimism about the future of health care in the United States. From the ranks of this progressive minority, the fund drew many members of its administrative staff, and trained their collective attention on health problems in the coalfields. These administrators and certain practicing physicians affiliated with the fund brought a concern for public health, a critical perspective on the living and working conditions of coal miners, and, in some cases, an explicit interest in the occupational origins of disease.

A new medical perspective on black lung, one that rejected the banality of "miners' asthma" and the narrowness of an exclusive focus on silica, began to emerge. Conferences, research projects, and articles in professional journals, virtually all of them originating with fund-affiliated clinics and physicians, evidenced the new attention. The new perspective asserted that coal dust without silica was indeed an agent of disease, and that it caused a specific clinical entity termed coal workers' pneumoconiosis (CWP). In 1954, a physician from Elkins, West Virginia, fired the opening salvo in what would become the medical battle over black lung disease. Assuming a polemical stance, Dr. Joseph E. Martin, Jr., argued in the *American Journal of Public Health* that "authoritative opinion to the contrary notwithstanding," coal miners suffer from a "disabling, progressive, killing disease which is related to exposure to coal dust."[56] Martin based his conclusions on a study by the British Medical Research Council and his own five years of clinical observations of 400 bituminous coal miners. Martin's article was only the first. Reports of clinical studies with similar conclusions soon began to appear in the professional literature.[57]

The staff of the Welfare and Retirement Fund intervened directly to promote recognition of coal workers' pneumoconiosis by the medical profession. In 1951, the fund's medical director transferred Dr. Lorin Kerr from a field office in Morgantown, West Virginia, to the headquarters in Washington, D.C., and directed him to develop expertise in the dust diseases affecting coal miners.[58] Fund revenues subsequently subsidized conferences on coal workers' pneumoconiosis and speaking tours for British doctors who had researched the disease. The medical staff also attempted to pressure the U.S. Public Health Service into conducting a major prevalence study of CWP, which they believed would inalterably establish the scientific credibility of the disease. Despite its own survey of coal miners' health records, which concluded in 1952 that "the rate of chronic chest disease was alarmingly high in American bituminous miners,"[59] the Public Health Service initially refused to undertake further investigation.

The persistence of the fund's medical staff, combined with the escalating attention CWP was receiving in U.S. medical journals, eventually yielded two major epidemiological studies of the disease. In 1959, the Pennsylvania Department of Health, in cooperation with the medical staff at certain fund-affiliated clinics, initiated what was at the time the largest survey of coal workers' pneumoconiosis in the United States.[60] Researchers collected data, including medical and employment histories, X-rays, and vital capacity determinations, on 16,000 working and retired bituminous coal miners in central and western Pennsylvania. Overall prevalence of the disease varied from 34 percent in central Pennsylvania to 14 percent in the western part of the state. The incidence increased sharply with age and number of years worked in the mines; among retired central Pennsylvania miners 65 years of age and older, 61 percent exhibited X-ray evidence of penumoconiosis.[61] Following publication of the initial findings from this study, U.S. Public Health Service officials also began to investigate CWP among bituminous coal miners. In a medical survey of over three thousand working and nonworking miners (the latter group included the retired, the disabled, and the unemployed), researchers found pneumoconiosis in 9.8 percent of the working miners and 18.2 percent of the nonworking group. The incidence of disease again rose dramatically with age, years of coal mine employment, and average dust exposure on the job most frequently performed.[62]

It would seem at this historical point in the early 1960s that black lung disease was virtually assured of recognition by the medical profession as a whole. That was not the case. "Authoritative opinion" persisted in ignoring and, in some cases, denying the existence of widespread disabling occupational lung disease among coal miners. Moreover, imbedded within the emerging research and literature on black lung were two conflicting definitions of disease, neither of which would achieve legitimacy solely by virtue of superior logic or scientific rigor. One perspective viewed black lung through the prism of scientific medicine's classical approach to the body and disease. Black lung was coal workers' pneumoconiosis, a disease associated with the inhalation of one specific agent, coal dust, which acted over time in an essentially mechanical and cumulative fashion to impair lung function. *Pneumoconiosis* is a generic term meaning "dust-containing lung"; by definition, the disease had to be diagnosed and classified through X-rays, for it was conceptualized in separate stages that were differentiated according to X-ray findings. The disease process, as constructed with this diagnostic tool, was linear and quantitative; the greater the size and number of opacities revealed on a chest film, the more advanced the disease.

A divergent view of black lung, more uncertain and speculative, yet more grounded in the physical condition of miners themselves, appeared during the same period. This conceptualization of disease took as its starting point the wide discrepancy between X-ray findings of pneumoconiosis and the actual respiratory impairment of coal miners. Chest X-rays of miners who suffered severe breathlessness and whose tests of lung function showed significant impairment did not necessarily reveal an advanced stage of pneumoconiosis; in some cases, they revealed little pneumoconiosis at all. A small minority of physicians began to investigate the possibility that these seemingly incongruous signs and symptoms of disease originated in the workplace. In 1964, one such group published its findings in an article on "Respiratory Disease in Southern West Virginia Coal Miners."[63] Financed by the U.S. Public Health Service and the Miners' Memorial Hospital Association, Drs. Hyatt, Kistin, and Mahan collected data on pulmonary function, incidence of pneumoconiosis, and respiratory symptoms in a group of 267 working and nonworking coal miners. The physicians then pursued the interrelationships between these medical data and variables like age, smoking habits, and work history. They discovered a direct relationship between impairment of a miner's lung function and the number of years spent working underground, regardless of age, smoking habits, or X-ray category of CWP. In conclusion, Hyatt and his colleagues argued: "Since progressive impairment of pulmonary function occurs in relation to years of underground mining even in the absence of pneumoconiosis, . . . it may be that harmful underground agents other than dust are responsible or that dust affects the lungs in ways other than by producing pneumoconiosis."[64]

Other physicians delved into an additional layer of complexity after observing that many of their coal miner patients, despite negative chest X-rays *and* normal pulmonary function tests, were nevertheless distressed by extreme shortness of breath. Prior research had tested only one aspect of miners' lung function: the ability to move air in and out of the lungs. Now physicians, Dr. Donald Rasmussen in particular, began to explore a second aspect of lung function: the ability actually to transfer oxygen into the blood. In order to test miners for this impairment, Rasmussen began putting his breathless patients through a time-consuming and exhausting blood gas analysis and "found very quickly that some of these coal miners with these symptoms did indeed have significant abnormalities in their ability to oxygenate their blood." Furthermore, "this occurs so commonly among our miners that we have no alternative but to accept it as an occupation-related abnormality."[65]

These different courses of research in effect focused on three distinct types of lung impairment: tissue destruction associated with pneumoconiosis; airway obstruction associated with bronchitis and other diseases; and oxygen deficiencies related to pulmonary vascular problems. To this day, these divergent approaches have not yielded a medical consensus on black lung. The confusing diversity of opinion raises the possibility that black lung is, in fact, several diseases, some or all of which may be associated with or aggravated by occupational exposure to dust (and perhaps other substances). At a deeper level, the medical disagreement reflects physicians' struggles to conceptualize a complex disease, or a complex of diseases, within a scientific paradigm that increasingly seems to have limited application and explanatory power. The scientific model of disease causation explicates the biological relationship between a specific agent and a distinct disease, and it has proved valuable in illuminating those illnesses that are compatible with this conceptualization. It does not, however, comprehend more complex interactions among numerous agents and simultaneous diseases.

Throughout much of the 1960s, the problematic nature of black lung disease remained the substance of abstruse conversations and research among a small group of physicians. The medical profession as a whole did not accept even the narrow definition of black lung as coal workers' pneumoconiosis. Some coalfield physicians continued to reassure their patients with the perverse homily "Coal Dust Is Good for You."* Others persisted in viewing miners' respiratory disability as a neurotic "miners' syndrome" "characterized by numerous somatic complaints, a passive, dependent attitude, and an outstanding lack of anxiety, with rationalizations based upon having been exposed to 'bad air' and hard work, and 'nerves being run down.'"[66] Such views died hard. Despite two large epidemiological studies by reputable agencies, the educational efforts of the fund's medical staff, and the clinical research of fund-affiliated physicians, the existence of a "new" occupational lung disease among coal miners was not accepted. Despite five decades of British research into coal workers' pneumoconiosis, and the extension in 1943 of compensation to its victims, coal miners in most areas of the United States received no workers' compensation for disability due to CWP. The recognition that occupational lung disease was rampant among coal miners did not evolve of its own accord within the boundaries of medical science. It was forced on the medical establishment through the decidedly political actions of coal miners themselves.

* This apparently derived from the perception that coal miners had a relatively low mortality rate from tuberculosis, considered a prime public health problem around the turn of the century. Inhalation of coal dust as a therapy for tuberculosis received enough attention to prompt one critic to argue against it in an editorial, "The Coal Mine as a Tuberculosis Sanitarium."[67]

# 2

## Where Is the Disease?

In the coal camps of southern West Virginia, the black dust of the mines is everywhere. It collects at the side of the narrow, two-lane roads, coats the green grass of lawns, and settles on children's toys left piled in the yards. Dust grays the sheets hung out to dry in the sun; it dulls the paint of houses, cars, and fence posts. When coal is cut from the seam where it has laid for eons, dust boils up like a thick soup and overflows onto every nearby surface. Even after a vigorous scrub in the shower, many miners emerge with the startling eyes of heavily made-up women, the mascara of the mines still clinging to their eyelids and lashes. Dust is a mundane thing, "fine dry pulverized particles of matter,"[1] but it is also a symbol. In the mining camps of southern West Virginia, where "Coal Is Our Life," dust is a symbol for the pervasive and intrusive presence of the coal industry.

From the perspective of science and engineering this dust is the cause of occupational respiratory disease; human interaction in the mines where dust is produced is of little relevance. The isolated physical object—dust—and the isolated biological object—the respiratory system—are the primary focus of scientific research, analysis, and intervention. And yet, within a deeper meaning of causation, this scientific perspective appears profoundly ideological, even ironically magical, for it robs human beings of their proper powers of action and creation and invests these powers in lifeless things, inanimate objects. Dust does not arise like a ghost in the night from the underground graves of the mines. Dust and disease are produced—no less than coal itself—through the agency of human beings in the workplaces of the coal industry.

In the capitalist economy of the United States, the central human relationship that has long structured the production of coal and disease is that between miners, those who actually retrieve coal from the earth, and operators (owners and their representatives), those who lay private claim to coal and to the apparatus of production. The antagonistic interests that are built into this class relationship have special implications for health and safety in the workplace. For miners, occupational hazards represent direct threats to their lives and health. Their financial survival depends upon their labor underground, yet this labor can bring, not survival, but disability or death. For the operators, accidents, disease, and similar byproducts of mining are fundamentally irrelevant to their economic objective in production, except perhaps on those rare occasions when the labor supply is insufficient to replace diseased and disabled miners. Safety hazards in the workplace (i.e., conditions producing injuries and fatalities, as distinct from disease) become a consideration only insofar as they jeopardize a capital investment; floods, explosions, major collapses of roof, and the like represent the destruction of capital and therefore must be avoided. Miners have long recognized the priority granted the protection of capital in the mines, as opposed to their own protection, and have expressed it in an old and common aphorism: "Coal operators always cared more for the mules [used for underground haulage] than the men. A man they just paid for the work he did; a mule they had to *buy*."

Within the limitations of this class structure, their actions have continuously shaped and redefined the relationship between miners and operators; the resultant changes often have carried important consequences for occupational safety and health. For example, extraordinary competition once surrounded bituminous coal mining and exerted relentless pressure on the operators to reduce the costs of production, regardless of the implications for working conditions or occupational hazards. Certain of the larger companies eventually were able to transform the fragmented and chaotic industry into a more concentrated and capital-intensive structure, but their methods of merger and mechanization produced negative consequences for working conditions underground. Miners, for their part, historically attempted to pit a countervailing power of resistance and organization against that of private property. Most importantly, they found collective strength in unionization, which made it possible for them to negotiate as a group the terms and conditions under which they would sell their capacity to work. Unionization represented an economic lever with which miners could influence the quality of their working conditions, but at times it also brought a new form of control and inhibition to rank-and-file power. These and other historical changes have influenced the production of occupational disease by altering the character of industrial relations in the coal industry. In certain periods, miners' unity and economic leverage have enabled them to enforce specific changes in the workplace, including dust reduction and other improvements pertinent to health and safety. In other periods, their economic and organizational power has faltered, with deteriorating working conditions the consequence.

Not only is the workplace production of disease contingent upon these relations of power; it is also the outcome of the actual organization and practice of work—the labor process—in a given industry at a specific time. The physical character of work clearly is a relevent factor in occupational safety and health, determining the type of danger that a worker may encounter. Coal miners are exposed to occupational hazards different from, for example, those confronting bakers, farmers, or chemical workers. The miners of previous generations, who hewed rock with picks and blasted coal with powder, encountered risks different from the risks of those who work in the mechanized and electrified mines of today. Moreover, at issue in the labor process is not simply physical surroundings, but also social organization: miners' authority over the activities and conditions of their work including certain aspects of their own safety and health changed dramatically with the transition from a craft to an industrial process of coal production.[2]

The organization and practice of work are not purely the results of physical or technological imperatives; they, too, are shaped by social and economic forces. In bituminous coal mining, the industrial relations between miners and operators influenced the timing and character of technological innovation, the division of labor, the pace of work, and other features of the labor process. Economic competition, combined eventually with the institutional power of the United Mine Workers to establish a high daily wage, pressed the operators to seek ever more productive methods. Their innovations affected the nature and extent of occupational danger by altering not only the physical practice of work (e.g., through mechanization), but also the social organization of the workplace (e.g., through intensified supervision and the erosion of miners' control over their craft).

Thus, understanding the origin of black lung—the fundamental cause of this disease—requires a historical examination of the changing labor process and industrial (or class) relations in bituminous coal. This history is played out in the workshops of mining engineers, on the picket lines of union organizing drives, and in the board rooms of coal corporations; its most important locus for the history of black lung, however, is the coal mine workplace, the site of disease production. Here, class relations intersect with the labor process to realize or reduce the potential harms of coal mining. Here, miners and operators exert their relative will and power to determine the actual conditions of work. It is in the historical context of this ever-changing physical site and social relationship that black lung disease is produced.

## "We Ask for a Mouthful of Fresh Air": Black Lung in the Hand-Loading Era

The work of underground coal mining first developed in the United States as a skilled craft occupation.[3] Like iron molders, glass blowers, and many other nineteenth-century workers, coal miners performed an entire sequence of tasks necessary to create a specific product. They owned both the knowledge and the tools of their trade: information about the process of coal production, as well as picks, augers, powder, and other supplies were in their possession and control. Young men and boys learned the trade from experienced miners, frequently their fathers or other male relatives, during lengthy apprenticeships. Once trained, they carried out their jobs relatively free from direct supervision or interference by management. For some miners and labor advocates, this craft era was a golden age of freedom.[4] For many operators, however, the autonomy of these craft workers eventually became an intolerable impediment to predictable output, increased productivity, and a secure margin of profit.

The craft process of mining coal took place largely within a single "room," one of which was allotted to each worker.[5] This space typically was about twenty feet wide, was enclosed on three sides, and opened at one end onto the entry, a tunnel used for the transport of coal, workers, and supplies. As the miner removed coal at the opposite end of the room—that is, at the face, the exposed area of the seam—his room elongated and the distance from the face to the entry increased. The height of the room depended partly on the thickness of the seam; in low coal, it was only four or five feet, just high enough to permit a loaded coal car to pass from the face to the entry.

Extracting coal from the hard surface of the seam required a series of tasks, all of which were usually carried out by the individual miner. First came the work of undercutting, one of the most skilled, dangerous, and time-consuming activities in the labor process:

> The miner's cycle of work . . . begins with the job of undercutting the coal by hand. To do this he hacks away at the bottom of the seam with a short sharp pick, cutting a "V" four feet deep under the solid coal. When the cut is part way in, he places a short piece of timber, a sprag, under the edge of the coal so that the overhanging mass will not fall on him as he lies underneath and finishes the job.[6]

After undercutting, the miner utilized a large auger to drill a hole of specified length, width, and angle into the coal seam. Into the hole went a cylinder of black powder, and then a thin role of waxed paper, which acted as a fuse. After igniting the fuse, the miner ran for refuge and then waited while the explosion cracked up the seam and released a ton or more of coal. The final step in the process was to load the coal onto a car: "Loading was a straightforward job of shoveling coal into the cars, though it was no simple matter to shovel coal in a tunnel rarely more than five feet high, and to throw it high enough to clear the sides of the car without hitting the roof or spilling over the far side."[7]

Miners also had to perform diverse preparatory and maintenance tasks, commonly called "deadwork" or "company" work because they were in effect unpaid tasks that were ancillary to production. Each miner was responsible for placing timbers to support the roof in his room. Laying track from the main entry into his work area was another responsibility. In low coal, it was often necessary to "take up bottom," or lower the floor of a room, in order to create sufficient height for the cars. Pushing two-thousand-pound coal cars up a slight grade to the face was especially arduous and sometimes resulted in back and other injuries. After performing all of these tasks, a miner in a good seam could still produce between four and five tons of coal during a ten-hour shift. At the turn of the century, he earned forty to fifty cents per ton for these labors, or about two dollars per day.[8]

The implications of this labor process for coal mine health and safety were mixed, suggesting limitations on the control miners actually had over their craft. It is true that miners worked largely without supervision and could control the pace and execution of specific tasks. Ideally, they were able to protect themselves from recognized dangers in their own work activities. However, miners' authority in the workplace ended at the entries to their rooms; it never extended to the infrastructure of the mine. With respect to the potential for occupational respiratory disease, this limitation was crucial, for miners did not control the essential mechanism of prevention: the ventilation system. Circulation of air was the central but not the only issue in ventilation. In some mines, where underground fires were used to draw air through the tunnels, the ventilation system itself was a hazard. Securing "a mouthful of fresh air" through statutory or collective bargaining requirements was a consistent health and safety goal of miners during this early period.[9] In the words of coal miner and union activist John Brophy: "Smoke from explosions of black powder, the reek of oil lamps, and the pervading coal dust made breathable air something of an obsession with the miner; . . . the miner had to pay for bad ventilation by 'miners' asthma' and other ailments."[10]

Ventilation tended to be poorest and respiratory hazards greatest in the nongassy mines. The prospect of losing their entire capital investment in one explosion encouraged mine owners to install better ventilation systems in mines where methane gas was present; the nongassy mines, however, tended to "kill the men by inches."[11] One mine inspector described the ventilation problem at the turn of the century and detailed the implications for miners' health:

> Adequate ventilation is not applied in such [nongassy] mines, because they can be wrought without going to the expense of providing costly and elaborate furnaces or fans, air-courses, stoppings, and brattice. From four to six cents a ton are thus saved in mining the coal that should be applied in ventilating, but saved at the expense of the workmen's health. . . . Constant labor in a badly-aired mine breaks down the constitution and clouds the intellect. The lungs become clogged up from inhaling coal dust, and from breathing noxious air; the body and limbs become stiff and sore, the mind loses the power of vigorous thought. After six years' labor in a badly ventilated mine—that is, a mine where a man with a good constitution may from habit be able to work every day for several years—the lungs begin to change to a bluish color. After twelve years they are black, and after twenty years they are densely black, not a vestige of natural color remaining, and are little better than carbon itself. The miner dies at thirty-five, of coal-miners' consumption.[12]

Further compromising miners' craft control and capacity for self-protection during this early era was the chronic economic instability of the coal industry. Prior to 1950, insiders as well as external observers frequently characterized bituminous coal as a "sick," "chaotic," and "unremunerative" industry.[13] Extraordinary levels of competition distinguished it from more oligarchic industries such as auto and steel. The geological blessing of abundant and widely dispersed bituminous coal deposits was an important factor in the competitive chaos that many operators came to curse. Entry into the industry was easy, requiring little capital outlay. In 1902, production was fragmented among an estimated 4,400 mine owners, none of whom controlled a significant portion of the national market.[14] Overproduction, excess capacity, price wars, and a small return on investment all plagued the industry until after World War II. Market booms, during which the operators received handsome prices for all the coal the miners could dig, only exacerbated economic instability over the long run; mines proliferated, production soared, and competition invariably sent the price structure into a tailspin.

These competitive dynamics affected miners in the form of irregular work, an unstable piece rate, and, in some cases, fraudulent weighing and screening practices. Because labor represented up to 75 percent of the cost of production, a prime tactic in the competitive struggle was to lower the cost of labor, principally by cutting the piece rate. Moreover, the craft nature of the labor process rendered companies relatively powerless to control miners' productivity, except indirectly by manipulating their wages. These economic pressures significantly circumscribed the "miners' freedom": a short work week or a cut in the piece rate forced them to work faster or work longer hours—or both—in an attempt to maintain their standard of living.

The impact on health and safety conditions was almost invariably negative, as miners necessarily reduced nonproductive, safety-oriented tasks, such as roof timbering, to a minimum.[15] Working longer hours in mines where "towards quitting time [the air] becomes so foul that the miners' lamps will no longer burn" no doubt increased the respiratory disease risk.[16] Moreover, a financially mandated speedup encouraged miners to reenter their work areas as soon as possible after blasting the coal loose from the face, an operation that generated clouds of dust and powder smoke.[17]

The hand-loading labor process thus carried the potential for specific types of injury and disease that the operators, pushed by the competitive dynamics of the industry, tended to bring to realization. However, coal miners did not passively accede to the manipulation of their wages and the erosion of their working conditions; they fought back principally by banding together in trade unions and presenting their own demands to the operators. Expansion of the United Mine Workers of America at the very end of the nineteenth century enhanced miners' influence over their working conditions; the union's development of its own institutional interests and policies, which in some cases collided with the desires of the rank and file, also affected workplace health and safety. The commitments and strategies of miners, operators, and the UMWA regarding occupational hazards differed in this period, but all of the commitments and strategies bore on the production of black lung disease.

## Strategies for Occupational Protection

The early history of organized efforts to reduce workplace dangers reveals the interplay of different interests and institutions in determining occupational health and safety.[18] Until the very end of the nineteenth century, unionization and the associated organized power of bituminous coal miners were at best localized and insecure. State law, as opposed to collective bargaining or federal legislation, was the first mechanism by which miners sought protection on the job; by 1900, an uneven collection of mine safety laws existed across the coalfields. Although generally ineffective and lacking meaningful enforcement provisions, these laws tended to vary in scope and detail according to the organizational and political power of coal miners in each state. The laws in Illinois and Pennsylvania, birthplaces of the first miners' unions in the United States and strongholds of the growing United Mine Workers, were generally acknowledged as the best in the nation; in West Virginia, largely nonunion at the time, the mine safety law was "the worst of the lot."[19] More importantly, accident statistics indicate that actual safety conditions underground also varied according to the extent of unionization: in 1907, there were 2.47

fatalities per 1,000 employed miners in the completely unionized states, 5.07 per 1,000 in those that were partially unionized, and 9.49 in the nonunionized states.[20]

Although it is true that the early controversies over state mine safety legislation generally featured a polarization of miners and operators, it would be wrong to conclude that profit-conscious operators always resisted safety precautions and that self-protective miners always supported them. The issue was far more complex. Indeed, coal operators were instrumental in the most innovative mine safety reform of the Progressive Era, the creation of the U.S. Bureau of Mines. The operators' interest in mine safety was concentrated almost exclusively on the single hazard that jeopardized capital investment as well as human life: explosions. This consideration informed their support for a federal bureau that would conduct research and education as to "the causes and conditions that have brought about these horrible explosions," as one operator put it.[21] Although roof falls, haulage accidents, and other sources of death underground year after year claimed more miners' lives than explosions, the Bureau of Mines' safety activities long reflected the operators' preoccupation with this one hazard.[22]

The industry broke with the precedent of state mine safety legislation and turned to the federal government during the Progressive Era in part because the diversity of state laws aggravated competitive instability; safety practices should be uniform throughout the industry, many operators argued, so as to eliminate them as a factor in competition. The operators did not, however, favor granting the federal government coercive powers to enforce safety in their workplaces. The U.S. Bureau of Mines was born in 1910, the offspring of these considerations. The enabling legislation and the bureau's subsequent activities reflected its conservative origins: the mine safety program focused on explosions and rescue operations to the exclusion of other safety and health hazards, and bureau officials were explicitly enjoined not to assume "any right or authority in connection with the inspection or supervision of mines."[23]

Miners, for their part, invariably included the protection of their lives and limbs among the formal goals of their early unions, including the United Mine Workers. Actual policies and actions with regard to occupational safety and health were variable and complex, however. Miners and their unions were generally the chief lobbyists for state mine safety laws; they also placed health and safety demands on their collective bargaining agendas. Although the interstate contracts first negotiated by the UMWA in 1898 did not explicitly cover health and safety, miners in the various districts appended clauses pertinent to ventilation, roof timbers (provided by the operators), and, in one case (Illinois), selection of the mine safety examiner by the workers.[24] Moreover, coal miners did not restrict their focus to occupational safety, which was a common limitation in the Progressive Era movements toward accident prevention and workers' compensation. Scattered references to "asthma" and "consumption" suggest that miners were aware of and disturbed by the extent of their own occupational respiratory disease; arguments for improved ventilation often included mention of this problem.[25]

Nevertheless, regulation of health and safety through legislation or collective bargaining was not a vital concern of the UMWA. The reasons for this were both economic and strategic. Miners faced a persistent and terrible contradiction between the need to earn a living wage and the need to protect themselves against dying in the effort. Immediate financial need often won out. Thus, the union fought safety laws that would have cut into miners' earnings, such as requirements to use more expensive but safer explosives. Feeding this opposition to legal restrictions on their work practices was the miners' desire to protect their craft control, as well as their pride in their own practical knowledge of the labor process. Contempt for "long-haired theorists" and "college professors" who presumed to dictate their work habits was especially evident during the political debate that preceded the establishment of the Bureau of Mines.[26]

Moreover, historical experience indicated that a legislative strategy did not produce positive results: not only were the laws ineffective, but the few people actually punished for violations were usually miners. In West Virginia, 98 percent of those prosecuted in 1910 under the mine safety law were workers; even in Pennsylvania, 80 percent of the prosecutions between 1908 and 1911 were against miners.[27] When inspectors brought charges against mine officials, local courts often refused to take any action. It is no wonder that one West Virginia mine inspector reported in 1907 that miners had "completely lost all confidence in the local courts . . . [and were] thoroughly convinced that justice could not be obtained towards the enforcement of the mining laws."[28]

Indeed, miners did not assume that legislation was the most effective means of achieving occupational safety and health, though historically that was the first mechanism to which they had turned. For many miners, occupational danger was one of numerous problems in the workplace, none of which could be solved without organizing collectively and altering miners' economic power relative to that of the coal operators. What this meant in practice varied according to individual political philosophy. Up through the 1930s and to a lesser extent after World War II, the left wing of the rank and file attributed their problems as workers, including occupational hazards, to nothing less than the private ownership of the coal industry. For the first three decades of this century, delegates to UMWA conventions articulated resolutions favoring various forms of public ownership; at the 1919 and 1921 conventions, members called for immediate nationalization of the mines.[29] A less radical though perhaps more common view emphasized unionization as the paramount strategy for protecting miners on the job. In this era of low wages, company towns, fragile collective bargaining, and rising nonunion coal production, many miners understandably saw industrywide unionization as the UMWA's major task. The previously cited accident statistics indicate that the connection between union membership and occupational safety was far more than rhetoric. As the UMW *Journal* once remarked concerning the high fatality rate in nonunion mines: "Death was due to lack of organization."[30]

The UMWA's early attempts to bring all miners into the union and negotiate industrywide contracts produced divergent results with regard to occupational health and safety. In some areas, the union was beaten back for decades, and miners' health and safety suffered as a consequence. In other locations, miners won the power of unionization and apparently achieved improvements in their working conditions; however, the pattern of industrial relations that was established between the UMWA and the operators would eventually place limits on the union's role in advocating for workplace health and safety.

In the southern coalfields, the operators' desperate need to minimize the costs of production and protect any competitive advantage generated a fierce and unyielding resistance to workers' attempts to organize. This was especially true at certain times and in certain places, such as in southern West Virginia during the 1910s and 1920s.[31] The bloody battles at Matewan, Paint Creek, Blair Mountain, and elsewhere were crucial episodes in coal miners' long efforts to unionize, and they have a special historical meaning to the black lung movement. Southern West Virginia was the birthplace of the controversy over black lung, and many miners who struck and demonstrated for recognition of the disease were veterans of these earlier battles. Utter devotion to unionism and a perception of the operators as ruthless enemies were the lifelong legacies of these armed confrontations; indeed, the legacies live on in a culture that transmits this history from generation to generation and in an economy that continues to pit miners and operators against each other.

Violent conflict did not characterize union organizing drives throughout the bituminous industry, however. Between 1898 and the end of World War I, many coal operators and unionized miners in Illinois, Indiana, Ohio, and western Pennsylvania were able to construct relatively peaceful—if sometimes uneasy—relations, based in part on the economic muscle of the UMWA in the region. Formed in 1890, the United Mine Workers remained small and ineffectual during its first seven years. In 1897, however, well over 100,000 workers responded to the call for a nationwide shutdown of the mines. Interstate collective bargaining ensued, and membership in the union expanded steadily through the northern fields. By 1904, 250,000 miners were UMWA members.[32]

Like their counterparts to the south, northern coal operators suffered financially from the excessively competitive structure of bituminous mining. Their responses to this problem contrasted markedly with the southern responses, however: rather than resist the UMWA at all costs, some northern operators anticipated the possibility of actually benefiting from industry-wide unionization. By standardizing the costs of production through union contracts, they hoped to reduce competitive instability and bring order out of chaos in the industry.[33] "Competitive equality" was a byword of the interstate collective bargaining conferences that took place between 1898 and 1908. What this meant was not a uniform piece rate throughout the unionized segment of the industry, but rather a complex system of wage differentials that took into account mining conditions in different areas. The goal was to equalize production costs, not wages.[34] In the operators' view, the union not only provided a mechanism for standardizing costs, reducing competition, and establishing a wage scale acceptable to miners; it also acted as an enforcer of the contract, disciplining miners who wildcatted or otherwise broke the agreement, and organizing the mines of recalcitrant operators who refused to bargain collectively. As one Pennsylvania mine owner wrote to UMWA president John Mitchell in 1904: "With honest, conservative men at the head of labor organizations the liability of having trouble is decreased and it is a safer method of settling wage questions than dealing with the rank and file of employees."[35]

The early period of cooperative collective bargaining could last only as long as a major portion of the industry participated. Nonunion coal produced in parts of Kentucky, West Virginia, and other southern states threatened the viability of the interstate conferences from the start. Repeatedly, northern operators urged the UMWA to unionize the southern fields; repeatedly, the union's organizing drives failed. By 1922, 64 percent of all coal was mined south of the Ohio River, most of it by nonunion miners.[36] World War I brought temporary prosperity and record production levels, but the crash that followed was only that much more severe in comparison. The price of coal plummeted, northern operators abrogated their contracts, strikes became lockouts, mines were reopened with nonunion labor. By 1930, UMWA president John L. Lewis presided over a decimated organization, bereft of members, money, and bargaining power. A decade of desperation began in 1920, but it eventually gave way to a new era in the history of the coalfields.

The short period of cooperation between northern operators and the UMWA leadership reveals an important theme in the union's institutional history that eventually had serious implications for the union's advocacy of workplace health and safety. Until the mid-1960s, extreme competition was the economic context for collective bargaining in the coal industry and provided a common enemy for coal operators and UMWA leaders to combat. Early on, the union's top officials were drawn into a collaborative relationship with certain operators that was based on their common need to reduce competition and stabilize the industry. This collaboration far exceeded joint campaigns to promote the domestic use of coal or boost the export market. It involved the conscious and pragmatic use of unionization and its primary tool, collective bargaining, to transform the economic structure of the industry. Implicit in such collaboration was a fundamental alteration in the meaning and function of unionism for workers: no longer a collective base of power from which to press rank-and-file demands, unionism became a means to standardize and enforce the costs of production industrywide. These tendencies were only incipient under UMWA president John Mitchell (1898–1908); they reached maturity under the command of John L. Lewis. Contradictions between rank-and-file desires for safer working conditions, expanded benefit programs, and other improvements and the economic needs of the industry inevitably arose. Thus, the rank-and-file discontent that eventually grew into an organized insurgency had roots far back in the history of the United Mine Workers and deep in the political economy of the bituminous coal industry.

## The Transformation of the Workplace

On December 14, 1948, the Joy Manufacturing Company of Pittsburgh, Pennsylvania, unveiled in a public demonstration the 3JCM-2 continuous mining machine, designed to rip two tons of coal per minute out of a solid seam.[37] The new machine represented the culmination of a gradual but profound transformation in the technology and social organization of the coal mine workplace. The change from hand tools to continuous process machinery, from craft autonomy to industrial discipline, and from an individualized work process to a complex division of labor required almost a century. Worker resistance, lack of capital, and, to a lesser extent, technical difficulties all impeded full mechanization of underground coal mining.[38] With the development of the continuous miner, full mechanization was at hand.

Early attempts to mechanize the labor process in coal mining focused on the highly skilled and time-consuming task of undercutting. In 1876, the Lechner Mining Company introduced the first cutting machine to coal operators in Ohio, and other companies soon began to market similar equipment. Initial adoption was uneven and was concentrated primarily in the unionized, higher wage fields of the North. Competition with nonunion, southern coal producers, especially for the lucrative Great Lakes market, pressed the northern operators to seek a more productive labor process. Hopeful that cutting machinery would boost productivity and lower their per unit costs, certain companies attempted to negotiate with the UMWA a "machine differential" (i.e., lower piece rate) that would justify their investment in the new equipment. Miners in some states—Kansas, for example—resisted the differential and thereby delayed the introduction of cutting machines, but in many areas the wage structure included allowances for the use of machinery. By 1900, one-third of Pennsylvania's coal production and nearly one-half of Ohio's was machine cut.[39]

In the nonunion mines to the south, mechanization came slightly later and for somewhat different reasons. Here, the surge in growth of the coal industry around the turn of the century set off a serious labor shortage. Coal operators responded by "de-skilling" the labor process, which enabled them to hire a larger number of inexperienced workers and to put them on the job after only a short period of training. Introduction of cutting machinery, combined with an increasingly specialized division of the labor process among machine operators ("runners"), hand loaders, shot firers, track men, and others, eased dependence on experienced miners. Machine mining relieved the operator "for the most part of skilled labor and of all the restraints which that implies. It open[ed] to him the whole labor market from which to recruit his force."[40]

Intensified supervision and the assertion of managerial control over the entire workplace sometimes accompanied these changes in technology and in the division of labor. At some mines, the operators focused simply on enforcing regular attendance by establishing strict policies regarding absenteeism, as well as a standard workday. (In those mines where miners could literally walk out, they traditionally controlled the length of their workdays.) In the mines of companies like U.S. Steel, however, already known for the "scientific" reorganization of its mills, the destruction of craft autonomy and the imposition of industrial discipline throughout the workplace was conscious policy.[41] By the early 1920s, workers at the company's Gary mines in southern West Virginia set timbers, laid track, installed electric lights, and performed other tasks according to engineering specifications transmitted by the foremen. Production schedules established on a daily basis set the pace of work. Foremen were responsible for meeting the specified quota of tonnage and for ensuring the safety of the approximately twenty-five workers under their supervision. Under this system the content of a miner's job did not necessarily change, but an essential element of his skilled status—the power to conceptualize and to control accordingly the execution of tasks—was removed. One of the company's daily safety bulletins explicitly reminded workers that the foremen "are supposed to do the thinking for the men."[42]

Mechanization of the loading process, which proceeded apace after 1930, brought the craft era of coal mining irrevocably to an end. Machinery replaced hand labor, and the social organization of the workplace changed. "De-skilling," increased supervision, and a more complex division of labor, which in some cases had accompanied the introduction of cutting machines, now became standard practice. A crew of workers, each member responsible for a distinct piece of the labor process, replaced the solitary miner in his room. The knowledge and execution of coal mining were increasingly separated, as engineers came to dictate the work practices that miners themselves had once controlled. Supervision intensified, as companies attempted to recoup their investment in machinery by pushing labor productivity higher and higher. By 1950, more than two-thirds of underground coal production was loaded by machine.[43]

New hazards accompanied these changes in the physical character and social organization of work. As one of the industry's own journals acknowledged, cutting machinery had a "tendency to produce deafness in those operating it."[44] Loud noise from the machines also prevented rapid communication in the event of emergency and inhibited miners' ability to hear the "top working," the customary warning of a roof fall. Certain models of cutting machines vibrated violently and sometimes caused internal injuries to miners.[45] The increased use of electricity in a close, dark, frequently wet environment also brought new safety hazards.

And then there was the dust. Miners whose working careers spanned the hand-loading and mechanized eras nearly unanimously indicate that a tremendous increase in dust levels accompanied the introduction of new machinery. Quantification of the increase is not possible, because dust, like most other hazards, was not monitored. Where machines cut coal from the face, where they scooped it onto conveyors, where they transferred the broken lumps from one piece of equipment to another—in these and more locations, great volumes of dust reduced visibility and slowly crippled the lungs of those who breathed it. One miner recalled: "That's when the dust really started increasing, when they brought them loading machines in. Then we really had a dust problem. Companies brought in big, powerful machines and put less emphasis on getting air across the face, and no water on the equipment."[46]

It was not just the machinery that threatened miners' health, however, but also the new social organization of the workplace. Miners' power to protect themselves diminished along with their control over the pace and activities of work. Although slowdowns, sabotage of machinery, informal workers' rules (e.g., a break at a specified time during the shift) and other forms of resistance inhibited the supervisors' capacity to direct the work force, formal authority over the pace and execution of tasks shifted to the section foremen. They in turn were held accountable by their supervisors primarily for the productivity—rather than the safety or the health—of the crew in their section.

Specialization and the increased division of labor also damaged workers' health by intensifying their exposure to certain hazards. Miners whose work once encompassed a varied assortment of activities now performed the same tasks over and over again. This increased not only the strain on specific muscle groups, but also miners' exposure to certain risks. Runners and drill men, for example, who day in and day out operated machinery that cut into the coal face, were constantly surrounded by clouds of dust. "Drill Man Blues," written by West Virginia miner George Sizemore, illustrates this relationship between specialization and occupational disease. The song was recorded in 1940 by George Korson, who commented: "Recording George C. 'Curley' Sizemore's mine ballads was a sad experience for me. . . . There were frequent breaks in his singing as he paused for breath. He said new ballads form themselves in his mind as he drills but he cannot sing them because he would get a mouthful of dust if he parted his lips."[47]

DRILL MAN BLUES

I used to be a drill man,
Down at Old Parlee;

Drilling through slate and sand rock,
Till it got the best of me.

Rock dust has almost killed me,
It's turned me out in the rain;
For dust has settled on my lungs,
And causes me constant pain.

I can hear my hammer roarin',
As I lay down for my sleep;
For drilling is the job I love,
And this I will repeat.

It's killed two fellow workers,
Here at Old Parlee;
And now I've eaten so much dust, Lord,
That it's killin' me.

I'm thinkin' of poor drill men,
Away down in the mines,
Who from eating dust will end up
With a fate just like mine.[48]

Miners sought to challenge the negative consequences of mechanization primarily through their workplace organization, the United Mine Workers of America. By the mid-1930s, the union was once again a potent force in the coalfields; in 1934, the ranks of the UMWA stood at 400,000. Organizing victories notwithstanding, miners who attended the union's convention in that year came with an array of grievances associated with the recent introduction of loading machines. Those accustomed to handloading were quick to rename the Joy loaders "man killers" and to protest the unemployment, physical hardship placed on older miners, and occupational hazards that attended their use in the workplace. Delegates hotly debated a resolution demanding the removal of these machines from the mines, and the few who spoke against the resolution were nearly shouted down by the tumultuous convention. One miner argued:

> I heard one of the brothers say that they don't hire miners over forty years of age in their locality. I want to tell you brothers that there is no miner that can work in the mines under those conveyors [loading machines] and reach the age of forty. Those conveyors are man killers and I believe this convention should do its utmost to find some way whereby those conveyors will be abolished. . . . The young men after they work in the mine six or eight hours daily become sick, either getting asthma or some other sickness due to the dust of the conveyors and they can no longer perform their duty.[49]

Despite protests and resolutions, mechanization continued; indeed, it escalated. The development of continuous miners after World War II was a technological giant step toward a fully mechanized labor process. Continuous mining technology concentrated the essential knowledge and tasks of the craft miner in a single piece of equipment that theoretically could be operated without interruption. The machine combined the jobs of undercutting, drilling, blasting, and loading by clawing out the coal with electrically powered rotating drill bits and transporting it away from the face on a conveyor belt. Remaining tasks were abridged with the introduction of roof-bolting machines and electrically powered trackless haulage equipment. Although mechanical breakdowns and worker resistance often rendered its advertised two tons per minute output a pipe dream, the continuous miner was largely responsible for the enormous postwar rise in labor productivity, which soared from 6.8 tons per worker-day in 1950 to 19.9 tons per worker-day in 1969. By 1970, half of all deep-mined coal produced in the United States was cut with these machines.[50]

Larger coal companies introduced continuous miners after World War II as part of a desperate competitive battle with oil and natural gas. Although these fuels had been encroaching on coal's markets since the early decades of the twentieth century, the trend accelerated during World War II. In the ten years between 1940 and 1950, the proportion of all U.S. energy generated by bituminous coal dropped from close to 60 percent to approximately 45 percent. In the same period, petroleum's share rose from around 17 percent to almost 23 percent, and natural gas cornered nearly 23 percent of production, up from about 12 percent in 1940. The railroad market revealed a much more dramatic shift. By 1950, diesel fuel was aggressively pushing the coal-powered iron horse into the realm of history books and museums, having increased its share of the railroad fuel market from a mere 2 percent in 1940 to nearly 50 percent only ten years later.[51]

Introduction of continuous mining technology was central to the larger coal operators' strategic defense of their markets. By mechanizing the labor process, producers with substantial financial resources dramatically increased miners' output while simultaneously reducing the size of their work forces. This enabled them to keep coal prices down and hold the line against oil and natural gas. Smaller companies that lacked the capital to mechanize were left behind with bulky payrolls and a price structure that did not cover their costs. Bankruptcies occurred with increasing frequency during the 1950s, and the bankrupt companies were often acquired by one of the larger companies. Indeed, there is evidence of a coordinated movement by the big operators to restructure the coal industry through mechanization, mergers, and acquisitions.[52]

Miners who tried to throw the institutional weight of the United Mine Workers against mechanization were unsuccessful. Under John L. Lewis, the UMWA went beyond accepting mechanization—it aggressively promoted the trend. The union was by far the most significant vehicle for miners to influence work-related issues, and its refusal to support rank-and-file proposals concerning mechanization doomed them to oblivion and failure. The UMWA's role during the 1950s has frequently been identified with the person of John L. Lewis, whose absolute control over the union's internal workings made him the chief architect of its policies. Yet, the union's role during this period was also rooted in history. It stretched back to the days when John Mitchell pursued a policy of cooperation with the coal operators based on the apparently mutual interest of operators and miners in stabilizing the industry, and it reached forward to the time when rank-and-file insurgency would topple more contemporary collaborators from power. The UMWA's postwar neglect of health and safety issues contributed to another generation of coal miners disabled by accidents and respiratory disease; neglect of members' concerns for seniority, job security, working conditions, and other issues scattered discontent throughout the coalfields, and created fertile conditions for the eventual growth of the black lung movement.

## In Like a Lion and Out Like a Lamb: The UMWA, 1920–1960

John L. Lewis's reign as president of the United Mine Workers of America included forty of the most important years in U.S. labor history, from 1920 to 1960.[53] He retained his grip on the union's helm during tumultuous internecine strife in the 1920s, emerging as the undisputed leader, and then presided over the union's renaissance during the 1930s. With the UMWA as his base, Lewis launched the Congress for Industrial Organization (CIO), placed himself at the forefront of the industrial union movement, and reached the zenith of his career as "Mr. Organized Labor."[54] By the time he retired as UMWA president in 1960, Lewis had acquired a mystique as "the greatest labor leader of all time"[55] and, on the more mundane side, had built an autocratic union structure, both of which guaranteed him supreme authority over UMWA policies and internal affairs.

Lewis was a visionary, a man who perceived clearly the potential power of organized labor and acted without hesitation at a strategic moment in history when that power was realizable. He was also a ruthless pragmatist.[56] As UMWA president, Lewis instituted policies that were informed by visions of an institutionally powerful union, a well-paid work force, and a thriving, technologically advanced industry. These goals were to be achieved through a single, long-range strategy. A powerful and centralized United Mine Workers could impose a high wage rate and more uniform working conditions throughout the industry. The high standard would squeeze the smaller operators into bankruptcy and force the larger ones to mechanize. This, in turn, would produce a streamlined, economically concentrated coal industry, which was a prerequisite for miners' (and their union's) long-range financial security and institutional power.[57]

Lewis first attempted to pursue this strategy during the 1920s, when he purged all rank-and-file opponents and centralized the union's power in his own hands. Simultaneous efforts to impose a high and uniform wage and to force the operators to mechanize met with less success. Transformation of the industry hinged on two requirements, neither of which was in place: first, industrywide unionization and collective bargaining, the vehicles for standardizing the wage rate; and, second, a sufficient concentration of capital to enable the largest producers to mechanize. During the 1920s, the elements of Lewis's strategy collapsed one by one as the nonunion companies ferociously resisted unionization, the companies in the Central Competitive Field abrogated their contracts and went nonunion, and membership in the UMWA shrank to one-fifth of its wartime level. By 1926, the three-year Jacksonville wage scale of 1924 was in force only in Illinois, and when that wage scale expired in 1927, Lewis authorized the district presidents to sign almost any contract they could get.[58]

Two decades later, the situation in the bituminous coal industry was altogether different: the work force was thoroughly unionized, and capital was more concentrated. During the 1930s, miners rebuilt the UMWA and extended its influence throughout the coalfields. Widespread organizing drives were initiated after passage of the National Industrial Recovery Act in 1933 and brought new members in droves. Miners from southern Appalachia tended to associate their long-sought victory with the person of John L. Lewis, and they provided him a formidable bloc of support within the United Mine Workers. The resurgent miners' union supplied key organizational and financial assistance to the great organizing drives in basic industry and provided the base from which Lewis could challenge the leadership of the American Federation of Labor and found the CIO. Even during World War II, when much of organized labor abided by a no-strike pledge, miners retained their militant stance and repeatedly shut down the coal industry. The fruits of these struggles were the closed shop in the captive* mines, portal-to-portal pay, higher wages, a federal mine safety code in the UMWA contract, and, most importantly, establishment of the Welfare and Retirement Fund.[59]

The first telltale sign of a new order in the bituminous coal industry was the formation of Pittsburgh-Consolidation Coal Company (Pitt-Consol) in 1945. This corporation, a merger of the Rockefeller, Hanna, and Mellon coal investments, was by far the largest commercial producer in the United States. At the helm of Pitt-Consol was George Hutchinson Love, a man whose economic strategy for the coal business echoed that of John L. Lewis:

> We felt that if we could eventually build a company that could afford to close down the poorer properties and concentrate on the better ones, there was a chance to make something out of this industry. And that is just what we were able to do by acquiring coal properties through purchase or merger.
>
> And there was no technology. Nobody could afford it. No company was large enough to work on it. When we finally got the present Consolidation Coal Company [the name Pitt-Consol ac-

*A captive mine is one whose output is consumed entirely by its owner, usually a steel company or utility, and therefore never reaches the commercial market.

> quired in 1958] together in the mid-40s, it had the assets and personnel to develop technology and do research. . . . If coal was going to be competitive with other fuels, like oil and gas, you had no choice but to mechanize.[60]

Soon after the emergence of Pitt-Consol, the Bituminous Coal Operators Association (BCOA) was organized. This group included the largest soft coal producers in the United States and soon came to represent about 50 percent of total national coal output, dwarfing the small operators' Southern Coal Producers Association by a factor of two to one.[61] Behind the organization of the BCOA was George Love, who served as chairperson only long enough to oversee the election of U.S. Steel's Harry Moses as president.[62] These two men represented an alliance between the steel interests and large commercial coal producers that was the backbone of the new association. The BCOA was designed to represent the large commercial and captive mine-owning companies during collective bargaining sessions with the UMWA; by speaking with one voice, BCOA members hoped to stabilize the negotiating process, augment their bargaining leverage, and present an organized counterbalance to the small operators' association.

The emergence of a corporate leader, Pitt-Consol, and the organization of the larger producers into a national trade association signified the growing concentration of capital in the coal industry. The largest coal companies, benefiting from their own wartime profits as well as postwar bankruptcies among the smaller operators, had finally acquired sufficient financial power to mechanize fully the labor process and to reorganize the economic structure of coal mining. Detente with the UMWA leadership, based on expectations of the mutual benefits to be derived from mechanization and consolidation, would soon take place during collective bargaining.

The wage agreement negotiated in 1950 now stands as a historical landmark in relations between the United Mine

Workers as an institution and the larger corporations in the U.S. coal industry.[63] For almost twenty years following the signing of this contract, labor relations were relatively quiescent in the unionized segments of the soft coal industry. The smaller operators, represented by the Southern Coal Producers Association, "remained on the sidelines," even though the same contract was for the first time imposed on the entire bituminous industry.[64]

The settlement embodied a conspicuous trade-off between the miners' right to strike and control of the Welfare and Retirement Fund.[65] The large operators, represented by Love, agreed to permit Josephine Roche, fund director and close personal friend of John L. Lewis, to sit as "neutral" trustee on the fund's tripartite board; this gave the union (i.e., Lewis) control of the fund. In return, Lewis erased from the contract the "able and willing" clause. Negotiated in 1947 as a contractual protection against the repressive provisions of the Taft-Hartley Act, this clause in effect gave miners the right to strike. It read: "This Agreement . . . shall cover the employment of persons employed in the bituminous coal mines covered by this Agreement during such time as such persons are able and willing to work."[66] In addition, Lewis agreed to confine what had been the right to unlimited memorial periods, or work stoppages in commemoration of mine accident victims, to five days per year.

For the BCOA, elimination of the miners' contractual right to strike was far more than a legal formality. Throughout the 1940s, in wartime and after, massive strikes had upset the industry, and the federal government had seized temporary control of the mines four times. What rankled the operators about these strikes was not only the coal miners' defiant solidarity and formidable bargaining leverage, but also the fear that this unruly labor force would eventually provoke permanent federal seizure of their coal properties. Great Britain's nationalization of its mines in 1948, and the pervasive image of the U.S.

coal business as a "sick" industry, added to the operators' apprehension.[67] Moreover, during the coal slump of the late 1940s, Lewis began to function as a major-domo of the industry, arrogating to himself the power to set production schedules and work weeks. In 1949, UMWA miners shut down the industry for two weeks to reduce stockpiles and then, on Lewis's orders, began to work three-day weeks. For the operators, control of their mines by the union was as repugnant as control by the federal government. Forfeiture of the "able and willing" clause in the 1950 contract signaled Lewis's agreement that a private coal industry was preferable to a nationalized one, as well as his willingness to allow the operators a chance to put their house in order. George Love commented after the settlement:

> *The operators definitely established the right to control their production and their mining facilities.* The union asked for a cooperative administration of the welfare fund and we are giving it to them. We will help every way possible to make this huge fund a definite credit and benefit to the industry. However, the responsibility is squarely on the shoulders of the union and if it fails, the public and ourselves will look directly at the union.[68]

Missing from the 1950 wage agreement was an explicit reference to mechanization, consolidation, or the shared economic perspective that underlay the new industrial relationship. That the UMWA collaborated with the large operators to achieve these goals is evident, however, from its policies in subsequent years. For the next two decades, John L. Lewis and his successors utilized the union's institutional power and activity to promote the restructuring of the coal industry. Political lobbying, financial investments, collective bargaining—all reflected this purpose. When rank-and-file miners called for a slowing of mechanization, reduction of occupational hazards, or a return to union democracy—as they did repeatedly—Lewis used his prestige and authority as president to silence them.

During each collective bargaining session after 1950, negotiators from the UMWA and the BCOA attempted a delicate balancing act between conflicting considerations. On the one hand, the weakness of the market and the big operators' huge capital outlays for machinery militated against large wage increases and improvements in working conditions. On the other hand, the desire to squeeze the small operators by increasing their labor costs was an incentive to boost wages as fast as the large companies could replace miners with machines. The result was twenty years of hefty but sporadic wage increases accompanied by a pronounced deterioration in underground miners' working conditions.

Although collective bargaining was the most significant and wide-ranging vehicle for influencing the economics of the coal business, other union functions were informed by the same goal. The UMWA's only serious advocacy of occupational safety during this period focused on extending the 1952 Mine Safety Act to include operators who employed fewer than 15 workers.[69] Passed by Congress after 119 miners in West Frankfort, Illinois, lost their lives in an underground explosion, the 1952 act granted the Bureau of Mines limited enforcement powers over explosion-related hazards. Enforceable provisions pertaining to roof falls, machine hazards, haulage accidents—that is, the greatest sources of death and injury—were absent. Although the act was generally deemed a "sham," even by President Harry S. Truman as he signed it into law, extending its coverage was the single-minded target of annual legislative campaigns by the UMWA.[70] The fatality rate in the small mines, which tended to be two to three times higher than that in the larger operations, provided a strong rationale for increased safety precautions.[71] Given the overall context of collaboration and union inaction on matters of underground working conditions, however, economic motives very likely exceeded safety considerations in the UMWA's lobbying. Federal safety regulations represented one additional cost of production to impose on the smaller operators. Perhaps the most telling comment on the union's role in occupational health and safety during this period came from BCOA president Harry Moses, who stated that the UMWA "joined with us without reservation in all our efforts to combat the influences of competitive fuels, government interference and unreasonable safety regulations."[72]

Even the Welfare and Retirement Fund, vital and unique though it remained, was circumscribed by the UMWA's policy of industrial collaboration. Both the scope of its benefit programs and the activities of its staff were checked by the priority given to the economic needs of the larger operators. After securing BCOA agreement in 1952 to royalty payments (the fund's source of revenue) of forty cents per ton, Lewis demanded no further increase. UMWA presidents Thomas Kennedy and William Anthony ("Tony") Boyle followed suit, and from 1952 to 1971, the royalty did not rise. As a result, the fund's programs were pared away, bit by bit. Benefits were eliminated, eligibility rules tightened, pensions cut, and the hospitals sold.

The trustees began restricting eligibility for benefits in 1953. Although the fund initially gave pensions to miners who had worked for twenty years during any period, applicants now were required to have worked in the coal industry for twenty of the thirty years preceding the date of application. This rule change was especially hard on the old-timers who had entered the mines before 1923 and, due to disability or sporadic work, did not accumulate twenty years of service after 1923. Simultaneously, the trustees restricted eligibility for programs benefiting widows, dependents, and those seeking rehabilitation and maintenance aid.[73] One year later, cash maintenance payments for disabled miners and widows were eliminated.[74] Six years later, stagnating output and insufficient royalties forced the fund to cut pensions from $100 per month to $75. At the same time, miners who for the previous year had been unemployed or had worked in a nonunion mine were abruptly denied medical and hospital coverage. And in 1963, after operating them at a loss for several years, the fund sold its ten hospitals to the Presbyterian church.[75]

UMWA priorities also restricted the fund's role in promoting the recognition and prevention of black lung, making its history with regard to this problem ambivalent. On the one hand fund staff campaigned extensively in the medical community for recognition of coal workers' pneumoconiosis, and fund revenues financed clinics where physicians undertook essential epidemiological research that documented the existence of the disease. On the other hand, information about CWP was never communicated to the rank and file, and neither the fund nor the union officially tackled the most important aspect of the problem—prevention.

Fund director Josephine Roche explicitly prohibited her staff from becoming involved in any political efforts to gain improved workers' compensation coverage or occupational disease prevention programs.[76] Her rationale was that the fund's tripartite board and coal company financial base rendered inappropriate such controversial activities. The credibility of this argument fades in the face of the historical record: from 1950 on, the fund was openly dominated by John L. Lewis, who served as chairman of the trustees until his death on June 11, 1969. Roche, the "neutral" trustee, was never known to vote against Lewis, so he controlled two of the board's three votes.[77] More plausibly, Roche may have feared that political activities around the issue of miners' occupational health would generate demands for fund disability benefits, a financial liability that the fund's undernourished royalty revenues could not begin to meet.

UMWA policy affected the fund most directly and fundamentally through the tonnage royalty system whereby it was financed. The tie between revenues and coal output gave miners a stake in boosting production and productivity and reflected Lewis's goal of a concentrated, highly productive industry. Alternative financing mechanisms, such as hours worked or number of employees, were not compatible with the desired trends toward a mechanized labor process and trimmed-down work force; they would have yielded serious shortfalls during this period of irregular work weeks and wholesale layoffs. The tonnage system, however, also had its drawback: when coal production faltered, the fund's treasury shrank.

Analysis of Lewis's rationale for refusing to seek a royalty increase is inherently speculative; his reasons were never recorded. The policy seems consistent, however, with the overall ends-means logic of the UMWA-BCOA collaboration. Restructuring the coal industry involved cutting out excess miners (and potential fund beneficiaries) like so much dead wood. As fewer and fewer workers produced the same amount of coal—and eventually more—the ratio of fund revenues to beneficiaries would gradually rise. Had Lewis demanded an increased royalty on each ton of coal during the transition period of mechanization, he would have penalized precisely those companies that were raising productivity, capturing larger portions of the market, and upping their total output. These, of course, were the large producers, whose expansion Lewis wanted to promote, not undermine. Therefore, the royalty remained at forty cents per ton for nearly twenty years.

UMWA control of the fund's financial investments was another avenue of influence that had serious repercussions for beneficiaries.[78] In late 1948 or early 1949, the union bought a controlling interest in the National Bank of Washington and soon transferred to it the assets of the Welfare and Retirement Fund. Lewis proceeded to allow a substantial portion of fund revenues to lie fallow in non-interest-bearing checking accounts at the bank.

> In 1956, thirty million dollars, 23 percent of all the Fund's money, was unused. It dropped to fourteen million in the low year of 1961, but by 1966, when the industry was in full enjoyment of its new prosperity, the Fund had fifty million not in use. A year later, the idle cash rose to seventy-five million, an astonishing 44 percent of its total resources.[79]

These bulging accounts—which represented pensions and health care coverage for thousands of miners and their families—gave Lewis tremendous financial flexibility and muscle. He utilized them to hasten mechanization and consolidation by entering the coal business himself. A total of $17 million went to favored coal operators in the form of loans to purchase new mining machinery. Fund assets were also used to buy coal properties and stock in coal-burning utilities, investments designed to influence companies to purchase only UMWA-mined coal. In a court suit brought by disabled miners and widows against the fund, presiding judge Gerhard Gesell observed of these activities:

> This intimate relationship between the Union's financial and organizing activities and the utility investment activities of the trustees demonstrates that the Fund was acting primarily for the collateral benefit of the Union and the signatory operators in making most of its utility stock acquisitions. These activities present a clear case of self-dealing on the part of trustees Lewis and Schmidt [who represented the operators], and constituted a breach of trust.[80]

The UMWA's financial dealings with the assets of the Welfare and Retirement Fund were in many ways the most shameful of its policies during this period. Mechanization was an economic necessity for the survival of the coal industry, and unemployment an inevitable byproduct; it is at least arguable that the union's support for mechanization hastened and streamlined an inherently painful process. However, Lewis's schemes with the assets of the fund took the union on a prodigal adventure that was neither necessary nor successful. Those who suffered because of these financial deals were not coal operators, bank presidents, or union officials; they were the widowed, the disabled, and the retired. On June 30, 1968, the fund showed a net gain of $110 thousand on common stocks for which it had paid nearly $44.2 million over ten years.[81] This was not even enough to keep up with inflation. At the same time, the fund was losing approximately $3 million annually by allowing its assets to languish in non-interest-bearing checking accounts at the National Bank of Washington.[82] Meanwhile, retired miners were struggling to live on pensions of $75 per month, and those disabled before working twenty years in the mines received nothing at all.

## The Control of Rank-and-File Dissent

For two decades, Appalachia's people flowed out of the mountains, wave after wave of migrants. Forced to leave their homes by relentlessly high unemployment, these migrants sought in the North what they were so persistently refused in Appalachia—a job.[83] Some managed to make it to Detroit and were hired in the auto plants. Others stopped in Indianapolis, Cincinnati, or Cleveland and became janitors, secretaries, or factory workers. Many found no employment anywhere and went on relief. Between 1950 and 1970, well over three million Appalachians were exiled from their region.[84]

The massive exodus began to unravel the complex weave of kinship, community, and workplace ties that characterized Appalachian society. The extended family, the basic social unit of this rural area, was pulled apart; young people migrated north, whereas older residents tended to stay behind. Those who remained in the coalfields watched the communities they had grown up in decay and die, as employment opportunities shriveled and friends and relatives moved away, leaving their empty houses to crumble slowly from neglect. Miners worked two- and three-day weeks, withstood periodic shutdowns, and hoped that their mine would escape permanent closure. Those who were laid off sometimes opened "doghole" operations, using primitive tools to dig the coal out of abandoned mines. Others left their families in the mountains, boarded with relatives or rented a room in the city, and commuted home to familiar terrain at every opportunity. On Fridays and Sundays, their cars crowded the major highways between Kentucky and West Virginia and points farther north.

For the black population of the coalfields, the effects of this economic transformation were most extreme. Once indispensable to a young and growing industry plagued by labor shortages, black miners now bore the extra burden of unsparing race discrimination. Evidence supporting this charge abounds not only in statistics but also in the memories of black and white miners.[85] The words of one who lived through it, a black miner from southern West Virginia, summarize the history of this period:

> I got laid off in '53, in November, during the black exodus of the fifties. Everybody black was getting cut off; it was straight out discrimination. Now a lot of men had been cut off; many white men were cut off. But you take ten blacks out of a community, you done stripped the whole community. I mean everyone.
>
> That was all over West Virginia. Families was divided. People that had never basically been out of the state in their lives, a lot of 'em were going. A lot of 'em went to cities and places like that. Most of these blacks left. . . .
>
> I got three children. One boy worked in the coal mines a little while, then I told him the same thing that happened to me would happen to him. One boy went to college. One's in the service. I pretty well indoctrinated these boys about the mines, 'cause I don't see no long run future in it. The only future is black lung and between now and the black lung is being worked to death.[86]

This dislocation was directly attributable to the rapid postwar mechanization of the mines, which allowed companies to slash their work forces by as much as 75 percent. In 1950, over 400 thousand miners were required to retrieve from the earth the 516 million tons of coal produced in the United States; in 1969, production of 560 million tons required only 124 thousand miners.[87] Unopposed by the union, many operators utilized this period of labor surplus to reconstitute the work force. Layoffs disproportionately affected black miners, staunch rank-and-file leaders, and those with the early symptoms of respiratory disease or other disabilities. One black miner commented:

> It was the most blatant outrageous discrimination you ever seen . . . What they would do, see: I'm cutting coal with a machine. Six weeks or two months prior to a cutoff, they would transfer me from a machine to another job in the coal mines. They would transfer me to a different section. They transferred me from third right to a section that was pulling back. Then suddenly you'd get this notice—they would shut down the whole section. It would always be a lot of black people, and some white people they didn't want—drunks, people crippled up, people who made trouble. It came to where when they transferred a group of blacks to another section, we would always assume they were cutting off.
>
> After a while they began to get in the contract seniority spelled out a little bit different. But by then, they had laid off who they wanted to. They had gotten the work force they wanted.[88]

This situation seriously eroded the power of the rank and file at the local union level, a concern that underlay much of the unrest over seniority, job security, and unemployment. For example, at the 1956 UMWA convention, one miner stated: "A good union man today isn't wanted around the coal mines. I believe most of you gentlemen know what I am referring to. If you are a union man you are a radical, you don't want to listen. They keep harping on you about production and cost and everything."[89]

Miners, thus, lacked the economic and organizational leverage to influence the changes in their working conditions. They went to work every day in the most hazardous occupation in the United States without the protection of dependable union back-up, effective federal regulation, or the once customary power to protect themselves. For two decades, there had been gradual improvement in the safety record of underground mines; now, beginning in the early 1950s, the fatality rate began to rise slightly and the injury rate stagnated.[90] Absent from these statistics and ignored by both the operators and the union were the clouds of disease-breeding dust generated by the new machines. Their effect would be visible only two decades later, when a generation of breathless, disabled workers retired from the mines.

Individuals who tried to defend themselves on the job felt the shadow of unemployment cast over them: "Safety—hmph! A man would say anything and the next thing you know he was transferred to the hoot owl and then they run him off."[91] Local union officials who attempted to represent the rank-and-file concern for health and safety or other aspects of the changing working conditions were as expendable and unprotected as anyone else. "Fact about it, the company tried to make it hard on the safety committees, on all the local officers. Company would tell the safety committees if they found something wasn't supposed to be going on and they reported it, they'd try to fire 'em."[92]

Despite their economic vulnerability, coal miners repeatedly tried to halt the progress—or at least mitigate the impacts—of the new economic order in the coal industry. They contested management decisions on the job, staged wildcat strikes, and brought their grievances to successive UMWA conventions. Discontent focused on the interrelated transformation of their workplace and their union: miners challenged layoffs, speedup, intensification of supervision, and other changes underground; they also sought to reclaim the institutional power of the

UMWA through the demand for autonomy—that is, the right to elect their district officials. Throughout the 1950s and early 1960s, these rank-and-file protests were dispersed, disconnected, and largely ineffective; they took place against enormous odds. That they occurred at all is testimony to the depth of discontent.

Delegates attending the 1956 UMWA convention, for example, brought a host of grievances concerning seniority, layoffs, workplace health and safety, and other issues. Floor debate over the implications of mechanization was intense and often heated. Some delegates echoed the concerns voiced by other miners over loading machines twenty-two years earlier:

> They are putting coal moles [continuous miners] in our mines, and I hope they don't put them in anybody else's mines. We had one man die from the effects of that procedure. We had to give them a 15-minute shift. We have had any number who have had to get off because of health. It seems that someone forgot the miners who have [to operate] the moles. . . . He stands up there and inhales the fumes and the oil and the steam that is created by the heat from the mole. He doesn't get sufficient oxygen.[93]

Anger over the deterioration of working conditions underground surfaced repeatedly. For many employed miners, this was the central issue on which they were prepared to take a stand:

> We can't fight the battles that have been fought. All we can do is keep on continuing to fight the battles that we are faced with. Those battles face us in the form of mechanization, in the form of speed-up, in the form of the men having to take conditions that are much more dangerous in some respects than they were before. These are the conditions that the men are complaining about, that you hear every day when you go into the mine. And it is those conditions that they want to be rectified, that they want to see rectified. . . . The past strikes that we have had in four, or five years have all resulted in the matter of conditions.[94]

Some miners linked the deterioration of working conditions and the decline of rank-and-file power to the widespread layoffs. Unemployment emerged as the unifying issue of the convention. Delegates pressed for a variety of related protections: a strong seniority clause to replace the contractual provision allowing operators to lay off miners according to seniority by job classification rather than minewide; improved unemployment compensation; and enlarged Welfare and Retirement Fund benefits for disabled workers. Moreover, miners made their own long-range suggestions for ameliorating the situation, including a six-hour day, a lowered age for receipt of social security old-age benefits, liberalized pension eligibility standards that would enable miners to take early retirement, and a contractual clause requiring helpers on all machinery.

The UMWA's uncritical support of mechanization also rekindled the demand for autonomy, and the issue attracted some of the most spirited debates at UMWA conventions throughout the 1950s. The history of the autonomy issue stretches back to the 1920s, when Lewis established his authority throughout the UMWA in the midst of bitter internecine strife and a determined anti-union movement among the operators. Many miners from the centers of insurgent power, especially Illinois and western Pennsylvania, adhered firmly to the cause of internal democracy even after 1933, when Lewis's power within the union was secured. They were joined by miners from eastern Ohio and central to northern West Virginia, where valiant but hopeless strikes were waged in 1931 under the leadership of the Communist party's National Miners Union and the independent West Virginia Miners Union.[95] Throughout the next three decades, miners from these areas were a thorn in Lewis's side, unsuccessfully but stubbornly raising the banner of autonomy in successive UMWA conventions.

During the 1950s, Lewis handled the rank-and-file unrest over autonomy, unemployment, machinery hazards, and other issues through a combination of tactical finesse and autocratic power. He capitalized on the ideological diversity of those supporting autonomy by delivering thundering speeches about the political opportunism and financial mismanagement that supposedly attended internal democracy. Striking local unions he fined $300, and militant leaders he threatened with suspension.[96] At the volatile 1956 UMWA convention, Lewis shrewdly allowed the delegates to speak at length about the problems associated with mechanization. Any concrete recommendations for change he stonewalled by referring to the National Policy Committee, a body made up primarily of his own appointees. As usual, he gave a lengthy and eloquent speech defending his policies as president as well as the recent wage agreement. He then called for a motion: "Do any of you boys want to make a motion back there? I want to hear an affirmative motion on this contract. If you make a negative one, I won't recognize it."[97]

Despite convention protests, wildcat strikes, and other efforts, rank-and-file miners failed to alter the policies of their union or influence the changes in their workplace. With backing from the UMWA, they had taken on the coal operators and won, but they could not fight both the union and the industry at once. As a result, the operators were able to reorganize both the workplace and the work force at will, with little accountability to rank-and-file needs or demands. The intersection of mechanization and a collaborative structure of industrial relations exacted a terribly heavy toll of occupational respiratory disease among an entire generation of coal miners. Black lung was only one of many long-lasting products of this period, but it was one of the most lamentable.

In 1960, the year Lewis retired as UMWA president, there were few visible signs of organized unrest among coal miners, but before the close of the decade, a powerful rank-and-file movement rocked the United Mine Workers. In 1968, miners in southern West Virginia resolved to campaign for the medical and legal recognition of black lung and began a struggle that set them against their union leaders, the coal industry, and the medical establishment of the United States. This time, union officials could not control the rank-and-file dissent, and their failure signaled the end of the collaborative era in industrial relations. The miners' collective movement became a bid to reclaim the economic and institutional power denied them for nearly twenty years. Their goal of black lung compensation became a demand for retribution from the industry for the devastating human effects of its economic transformation.

# 3

# The Contagious Spread of Rebellion

Discontent hung like a storm cloud over the Appalachian coalfields. For nearly fifteen years, miners and other residents had endured a harsh economic depression. Most dealt with the deprivation in the privacy of their families—by cutting expenses, pooling resources with relatives, and commuting to areas where jobs were available. But with the beginning of a new decade in 1960, the political atmosphere began to change. Collective protest gradually replaced individual stoicism as the sanctioned response to hardship. Organizations based in both the workplace and the community began to coalesce, gain momentum, and win victories. The protests they launched were important forerunners to the black lung movement. They prefigured its constituency, issue, and organizing tactics. They transformed the political climate in the coalfields from acquiescence and resignation to rebellion and discontent.[1]

Protest originated from three distinct but overlapping groups of people who gradually became more visible, organized, and effective. Workers in the huge mines that lie in a wide circle around the adjoining borders of northern West Virginia, eastern Ohio, and southwestern Pennsylvania became the nucleus of rank-and-file discontent within the UMWA, which increasingly took the form of wildcat strikes, insurgent campaigns for union office, and challenges to the chair at union conventions. In southern West Virginia, a succession of disabled miners' and widows' organizations arose to dispute the Welfare and Retirement Fund's eligibility requirements for pensions and medical coverage. And in scattered locations throughout the southern coalfields, the war on poverty encouraged, through its finances and ideology, the formation of "poor people's" organizations that clashed with local elites over strip mining, political corruption, substandard public education, and the like.

All of the indigenous participants in these protests—whether widows, disabled miners, or welfare mothers—found their lives shaped by the production of coal in Appalachia. This partly explains the simultaneity of their political activity, for each participant was profoundly, although somewhat differently, affected by the structural changes in the postwar coal industry. Working coal miners bore the brunt of mechanization, speedup, occupational hazards, and the changed role of the union. Disabled miners, their wives, and miners' widows felt their already desperate medical and financial circumstances worsen when the Welfare and Retirement Fund began to cut off their benefits. The ranks of the poor (that is, those who not only have little money but also live on the margins of the wage economy)* multiplied rapidly, and their deprivation intensified during this period of high unemployment and a depressed coal market. Common structural origins lent a unifying historical meaning to their diverse protests. After more than a decade of regionwide dislocation that had substantially altered economic opportunities and the structure of local communities, miners, widows, and other residents sought to renegotiate the "social

* Disabled miners and widows are in some cases a subset of "poor people" as defined above. However, disabled miners have a history of coal mine employment and trade union affiliation that sets them apart from the subsistence farmers and others who are "poor"; in addition, their spouses or widows may be employed.

contract" with institutional elites. At the heart of their apparently separate, episodic battles was a common, underlying struggle over the new terms of class relations in the coalfields: Who would define the terms, and what would they be?

Economic depression provided the context and impetus for dissent, but additional factors made it possible.[2] During the early 1960s, key regional institutions began to waver in their capacity for social control. The retirement of John L. Lewis did not alter the formal powers of the union president, but no successor could dominate the UMWA with his authority, prestige, or success. No one could command equivalent loyalty from the rank and file. At the same time, the war on poverty weakened the authority of local and state political elites by placing funds and community organizers at the service of the disenfranchised poor. Even the economic power of the coal operators began to falter during this decade, when the Vietnam war siphoned off prospective miners into the military and a boom in the coal market simultaneously increased the industry's need for labor. Indeed, although the protest had its origins in the devastation of the 1950s and first emerged in the early 1960s, it only became effective and widespread in the relative prosperity at the end of the decade.[3]

Once set in motion, the rebellion gained momentum. Meetings of disabled miners, local unions, and community organizations turned into popular forums where anger and indignation could be aired. Scattered victories strengthened the belief that change was indeed possible. In this era, poverty and protest had a romantic allure, and external resources began to flow into the region; they came not only in the form of money, but also in the person of young organizers from the student and civil rights movements. By the late 1960s, there existed in the Appalachian coalfields an explosive mixture of deep-seated grievances, widely distributed community organizers, enhanced economic leverage among workers, and a record of recent popular victories. Each protest fed the spirit of solidarity and resistance.

The stage was set for a massive strike over black lung.

## The Human Toll of the Coal Industry: Disabled Miners and Widows

For those who must work to eat, to be permanently and completely disabled before the age of retirement is to confront the possibility of social, economic, and, sometimes, psychological ruin. It can mean not only alterations in physical appearance and associated social stigmas, but also the sudden elimination of a fundamental source of identity and self-worth: the capacity to perform financially remunerative work. It can mean facing before one's time many of the problems of the aged—high medical bills, a low fixed income, social isolation—yet, frequently, with the added responsibility and anxiety of providing for small children. For the spouses of the disabled, it can mean working a triple shift: as nurse, housekeeper, and family breadwinner.

Occupationally caused disability and even death are neither isolated nor infrequent occurrences in the U.S. coalfields. Virtually every family with a tradition of working in underground coal mines has lost the life or health of a brother, uncle, father, husband, or son. Some families have lost so many members that surviving a lifetime of work in the mines unharmed appears close to miraculous:

> I done my work at a coal mine in Raleigh County—Cranberry. It's worked out. I got my left leg hurt in June, 1941—June 2. They amputated it. And on June 6, 1972, I got the other one hurt. June ain't my month. . . .
>
> My daddy got killed there in the mines when I was two years old. My stepdad got killed there too, in the same mines. And Mom, she married a third time and he got hurt there, so he left and went to work at Skelton [and later succumbed to black lung disease].[4]

Recognizing the extraordinary level of economic need and

appalling health problems of workers who had received inadequate medical treatment for assorted occupational disabilities, the UMWA Welfare and Retirement Fund initially established several programs specifically for disabled miners. Paraplegics and others seriously disabled by mine accidents were transported by train at fund expense to medical centers throughout the United States, where they received physical therapy, prosthetic devices, and other appropriate treatment from medical specialists. Disabled miners and miners' widows who did not meet the eligibility standards for a pension of $100 per month received a stipend of $30 per month, plus $10 per month for each dependent.[5] Too young for social security old-age benefits and frequently ineligible for public assistance,[6] many disabled miners and widows came to expect and depend upon these benefit programs.

Humanitarianism and economic efficiency did not mix easily in the coalfields of the 1950s and 1960s, however. UMWA officials' single-minded promotion of a new economic structure in the coal industry soon compromised the generosity of the fund's benefits. Faced with recurrent fiscal crises, the trustees targeted for cuts programs that served those who by virtue of disability or irregular employment were economically marginal to the industry. Unlike employed miners—who were in effect paying for their own medical coverage and pensions through the coal they mined—the unemployed, the disabled, and the widowed represented a financial drain on the assets of the fund. Furthermore, they formed a relatively expendable group in terms of union politics: miners who were disabled or unemployed for a long period often stopped participating in the activities of their local union; miners' widows were not UMWA members and, except under unusual circumstances, had little influence in the union. For these reasons, but above all because of the need to relieve the fund's revenue crisis, programs for disabled miners and widows were curtailed during the 1950s and early 1960s. In 1954, financial stipends to more than fifty thousand disabled miners, widows, and children of deceased miners were eliminated. In 1960, the fund trustees ruled that after four years, permanently disabled miners would lose their "hospital cards," certificates of eligibility for medical and hospitalization coverage.[7] The situation was painfully ironic: men who while mining coal had suffered the priceless loss of their capacity to work were increasingly denied industry-financed benefits because they were no longer economically productive and able to work.

The contradiction between what they had lost to the coal industry and what they now received back from it became a bitter theme in the lives of many disabled miners and widows. An associated belief that they were *entitled* to benefits—their "just due"—became a central impulse in both the early mobilization around fund benefit programs and the later movement for black lung compensation. As the wife of one disabled miner put it:

> My husband killed his self in the mines. He led the weigh sheet in loading coal, but he killed his self. He's still living, but that's where his health went—to the mines and in the service. He gave it to his country in both ways. My husband had a rough life. That's why I think he should have every damn thing that belongs to him.[8]

Beliefs that the high-ranking officials of the fund and the union had neglected the needs of ordinary miners, and that democratic procedures should be instituted to force their accountability, also inspired the early organizing efforts of the disabled miners and widows. Ample pensions awaiting UMWA leaders deepened the sense of injustice: by 1967, miners able to meet the eligibility requirements for a pension received $115 per month, or $1,380 per year; UMWA president Tony Boyle and other officials enjoyed a separate pension fund, supported by union dues, that guaranteed the president a retirement income of $50,000 per year for life.[9] The arbitrariness of the fund's eligibility requirements and the absence of formal due process in its application procedures reinforced the desire for accountability. For example, some union miners who had worked their lifetime in the mines were denied a pension because mechanization and the UMWA contract's weak seniority clause left them unemployed; forced to work in nonunion "dogholes" in order to make a living, these miners were turned down because their last year of work had not been in a UMWA mine—one of several eligibility requirements for a pension. Others simply received denial letters, in which they found no comprehensible explanation for the fund's action, but did find the following statement affirming the absolute authority of the trustees: "This benefit is subject to suspension or termination at any time by the trustees of the fund for any matter, cause or thing, of which they shall be the sole judges, and without assignment of reason therefor."[10]

In order to force the fund's trustees to restore what "belonged" to them, disabled miners in southern West Virginia launched the first formal organization since the 1930s that was separate from the UMWA yet was designed to reform its policies.* On July 15, 1960, two weeks after the fund's ruling on hospital cards took effect, about three hundred disabled miners gathered in Glen Morgan, West Virginia, to protest the cutback and to discuss how to retrieve their benefits. The meeting was disorganized and tumultuous: speakers exhorted the crowd to "picket every coal mine in West Virginia"; angry miners shouted down the president of UMWA District 29 (southern West Virginia), George Titler, when he attempted to speak; and the man who appeared to be leading the meeting, a local union officer named Samuel Goddard, hurled sarcastic insults at the UMWA and fund leadership.[11]

* The disabled miners' groups (which later included widows in their name and membership) directed their demands at the fund, which was administratively and legally separate from the union. However, they were also extremely critical of the UMWA leadership, whom they (correctly) saw as dictating fund policies.

Over the next few days, miners articulated a range of political positions and tactics in response to the hospital card crisis. In nearby Logan County, they decided at a mass rally to send representatives to confer with fund officials in Washington, D.C., about the cutbacks. John L. Lewis and fund director Josephine Roche met personally with the Logan delegation, but the miners returned empty-handed. While this futile meeting was taking place, pickets protesting the fund ruling closed four mines in Logan County but managed to hold the men out for only a few days. Meanwhile, in Raleigh County, miners with a different point of view contested Goddard's leadership of the fledgling disabled miners' organization. It appears that at least some of the challengers considered him too antagonistic to UMWA and fund officials; they wanted to focus their attention on services and legislative reforms that would benefit disabled miners. This alternative group became known as the Disabled Miners Association, and it concentrated on obtaining prepaid hospitalization coverage through local medical facilities—a goal in which it was remarkably successful.[12]

The first angry protests over disabled miners' loss of medical coverage faded quickly. Within the year, Goddard's militant but unstable and ineffective organization passed from the scene. Although they left no visible impact on fund policies, these disabled miners made an unrecognized contribution to the development of rank-and-file struggles in the UMWA. They voiced the first public condemnations of the union leadership during the postwar period in southern West Virginia and thereby cut a path that others would follow into treacherous political terrain. Lewis's refusal to seek an increase in the operators' royalty payments combined with stagnating levels of national coal output to guarantee that further fund cutbacks, especially in pensions, would continue to generate dissent. Changes in the workplace, the union, and the coal industry were gradually creating the social base for insurgency, so that even when the fund began to augment its programs during coal's resurgence in the mid-1960s, rank-and-file rebellion simply continued to spread.

During the regionwide swell of political activity around strip mining, welfare rights, roads, public education, and other issues, in 1966 disabled miners and widows in southern West Virginia once again began to organize. Proclaiming that their purpose was to obtain pensions for disabled miners, widows, and dependent parents, the Association of Disabled Miners and Widows began meeting in the spring of 1967 at the Raleigh County Courthouse.[13] This "organization of living dead men trying to help each other" rapidly drew to its meetings several hundred miners, miners' wives, and widows from Raleigh and neighboring Boone, Fayette, and Wyoming counties.[14]

The Association of Disabled Miners and Widows differed in political strategy from the organizations that preceded it. Although focused on the Welfare and Retirement Fund, the association did not attempt to change the fund's policies through wildcat strikes or direct rank-and-file pressure on the union; rather, the group utilized court suits and congressional investigations to expose and challenge specific fund practices.[15] By January 1968, the group was pursuing a comprehensive list of legal demands, including not only technical changes in the eligibility standards for fund benefits, but also revisions in the silicosis provisions of the West Virginia workers' compensation statute.[16] By the end of February, the association had filed a total of four lawsuits against the Welfare and Retirement Fund and the West Virginia workers' compensation system. Suits against the fund charged that the trustees had acted "arbitrarily and capriciously" in deciding benefits eligibility, and suits against the compensation system challenged certain regulations governing occupational disease compensation. These cases were soon overshadowed, however, by the disabled miners' and widows' more publicized and significant class action lawsuit against the fund, known as Blankenship v. Boyle.

In August 1969, during Joseph ("Jock") Yablonski's unsuccessful campaign against Tony Boyle for the UMWA presidency, the Association of Disabled Miners and Widows filed a seventy-five-million-dollar lawsuit against the fund, the union, the BCOA, the National Bank of Washington, and certain individual UMWA and fund officials.[17] The suit charged: The "Trustees have violated their duties as Trustees for their own or others' profit and benefit, and they have exploited, made use of, and permitted the use of the assets of the Welfare Fund." Fund regulations were termed "arbitrary, capricious and unreasonable," and the trustees were accused of using their power to grant or withhold benefits "as a weapon of intimidation."[18] During almost two years of investigation, publicity, and litigation, this court suit became the singular reason for the association's existence. On April 28, 1971, the suit was decided in favor of the plaintiffs by U.S. District Court Judge Gerhard Gessell.[19] Having accomplished what had become their only goal, the disabled miners and widows declared a victory and disbanded their organization.

The disabled miners and widows of southern West Virginia were pioneers in the movements for black lung compensation and union democracy. In ideology, organization, and strategy, they anticipated other activists. The demand for improved workers' compensation coverage was the major focus of the black lung controversy; the beliefs in entitlement—"what's mine by right"—and in the accountability of a bureaucracy to its constituents were important to its ideological base. Disabled miners and widows also anticipated the Black Lung Association in the form and stance of their organization. As a community-based group, they united black and white and women and men around a common, work-related issue.

Moreover, disabled miners and widows played a courageous role in separating support for the union leadership from loyalty to the United Mine Workers as an institution. In southern West Virginia, union membership among coal miners is a basic symbol of personal integrity, and loyalty to the UMWA was tied historically to support for the policies of John L. Lewis, his successors, and appointees. Questioning either the union or its leaders could be an invitation to violent physical assault. The following story, told by a disabled miner's wife who was quite active in the Association of Disabled Miners and Widows, illustrates the emotional intensity of this issue:

> In 1943, I was working in Maryland and I went to Alexandria to see John L. Lewis' house. There was a fence all around and a guard standing there and he said, "You can't go in there. That's Mr. Lewis' house." And I said, "That's not *his* house. That house was built by coal miners all over the country." Well, I kept hollering and finally John L. Lewis and Josephine Roche came out and gave us a tour of the house.
>
> Now, my dad never hit me when I was growing up, but when I came back to West Virginia my dad was talking about John L. Lewis and I guess John L. Lewis was a big man to him. But he was not giving the miners credit for making the man, for making him so great. So I said, "You mean to tell me you're sitting here in this little four-room house and him living up there in a big house where I got mixed up just counting the bathrooms? He ain't nothing. He wouldn't have a damn thing without you all." And my daddy slapped my face. And I was married and had two children of my own.[20]

Finally, the early organizations of disabled miners and widows functioned as a training ground for some of the leaders and activists of more recent movements. Once the upsurge over black lung began in the early months of 1969, many members of the Association of Disabled Miners and Widows joined wholeheartedly in picketing, lobbying, and other political actions. Many subsequently devoted themselves to the Yablonski campaign for UMWA president, to passage of the 1969 federal Coal Mine Health and Safety Act, and to support for the continuing struggles over union reform and black lung compensation. Indeed, once the black lung strike of 1969 inaugurated a broad-based rebellion in the coalfields, distinctions among those in pursuit of reform began to collapse, as disabled miners and widows, black lung activists, and union insurgents joined together in widespread social upheaval.

## War on Poverty or War on the Poor?

The war on poverty was born amid racial strife and political compromise.[21] Begun only months after the historic "March on Washington for Jobs and Freedom" and the assassination of President John F. Kennedy, the poverty programs represented above all a response to the civil rights movement, which threatened not only to inflame urban ghettos and southern hamlets but also to tear apart the Democratic party.[22] Far from a head-on battle with systemic racism, however, the war on poverty was directed against the much more vague and general problem of the "poor." Structural economic issues like chronic unemployment, occupational segregation, and corporate control of regional resources not only were ignored, but, worse, were screened from view by the Johnson administration's appealing but deceptive rhetoric about a "Great Society" and an "unconditional war on poverty." At the same time, certain poverty programs became vehicles for establishing feisty poor people's organizations that demanded a genuine assault on inequality. The organizations were rewarded by having their funds cut off and being dissolved, but not before they skirmished repeatedly with political and economic elites, who feared that a new Reconstruction was in the making.

President Lyndon Johnson formally declared the war on poverty on January 8, 1964, in his State of the Union message.[23] Within a few months, the war was legislated into existence through the Economic Opportunity Act and ensconced in a new bureaucracy, the Office of Economic Opportunity (OEO). Chief strategists for the war tended to believe that the "enemy" was inadequate cultural and material resources, which left successive generations of poor people without the necessary skills and motivation to become economically productive. As President Johnson's Council of Economic Advisors observed in their 1964 annual report: "The chief reason for low rates of pay is low productivity, which in turn can reflect lack of education or training, physical or mental disability, or poor motivation."[24] Accordingly, programs like Job Corps, Upward Bound, and Head Start targeted aspects of the poor themselves that apparently needed revision: their education, job skills, health, value systems, and so on. This approach drew fire from both the Left and the Right. It was, some critics announced, a war *on* the poor.[25] Or, as Robert Sherrill observed in his comments on Job Corps, it was "just plain war": the program enabled participants to pass the Selective Service exams, so that more than one-fourth of the corps's first graduates went directly into the military to fight in the Johnson-escalated Vietnam War.[26] Critics on the Right directed their attacks primarily at the community action programs, which became the most controversial component in the antipoverty effort. Required by law to involve the "maximum feasible participation" of the poor, community action frequently evolved into a vehicle for political organizing; business elites and party regulars at city hall were understandably outraged by federally financed challenges to their power.[27]

Although many OEO programs served an urban black constituency, a national poverty effort by definition also included the poor of Appalachia. Poverty in the southern mountains was "rediscovered" in the course of the 1960 elections, when Protestant, working-class, economically depressed West Virginia became an important testing ground for the wealthy Catholic presidential candidate from Massachusetts. The Eisenhower administration's refusal to enact "depressed areas" legislation was viewed as a key reason for Republican congressional setbacks during the 1958 elections, and presidential candidate

John F. Kennedy took care to promise "a complete program to restore and revive the economy of West Virginia."[28] West Virginia's Democratic voters were never rewarded with comprehensive economic development, but Kennedy did promote scattered federal initiatives that benefited the poor of the state: a pilot food stamp program, as well as expansion of Aid to Families with Dependent Children (AFDC) to include children with unemployed parents, among others. In 1963, he established the President's Appalachian Regional Commission, which was designed to investigate and propose solutions to the problems of the area.[29]

By the time Lyndon Johnson was elected president, poverty had become fashionable, and OEO money soon was flowing into slums and hollows all over the country. Counties in South Carolina, New York, and Mississippi were thrown into the hodgepodge officially designated as "Appalachia," and the entire area was targeted to receive federal assistance in the Appalachian Regional Development Act of 1965.[30] This bill represented a "bricks-and-mortar" approach to development: building a more sophisticated infrastructure, especially roads, was considered the key to economic prosperity.[31] The act also created a new federal-state agency, the Appalachian Regional Commission, to oversee and coordinate economic development efforts.

There was a yawning gap between the promises and the programs of these national initiatives. They were dedicated to goals of prosperity and development, yet they failed to address the structural economic problems of central Appalachia: dependence on a single extractive industry, bituminous coal; concentration of land ownership, mineral wealth, and employment opportunities in absentee corporate hands; domination of the political process by coal interests, and the associated paucity of tax revenues for public education and other services. Local activists and organizers who genuinely sought to eradicate poverty increasingly realized that political change and economic development would come about only if they attacked these explosive and fundamental problems themselves. Community-based organizations committed to that purpose began to spring up throughout the mountains. Here, in local projects at the level of small, rural communities, the most significant impacts and confrontations of the war on poverty in Appalachia resulted.[32]

Among the most controversial OEO-funded programs in the region was the Appalachian Volunteers (AVs). Originally financed on a slim budget provided by the Council of the Southern Mountains, this effort began in 1964 as the summer work project of three students from Berea College who set out to renovate dilapidated public schoolhouses in eastern Kentucky. The orientation of the AVs soon moved from social service toward political organization, mirroring the evolution of many other community-based antipoverty efforts. In 1966, the U.S. Office of Economic Opportunity began to pump money into the program, and 500 AVs spread out through the hills of Kentucky, Tennessee, West Virginia, and Virginia. There they joined VISTAs (Volunteers in Service to America, a domestic version of the Peace Corps), community action staffers, and dedicated local activists in trying to build poor people's organizations.[33] In eastern Kentucky, the AVs turned from schoolhouses to strip mining; they joined with poor people in challenging the coal operators who were stripping away the hills, thereby creating landslides, stream siltation, and acid drainage. In Mingo County, West Virginia, they attempted to topple the "courthouse gang," as well as the traditions of vote buying, election fraud, and patronage, which dated back to the early days of the coal industry. Before an effective counterattack could be mounted, poor people throughout the southern Appalachians came together with the help of OEO-financed community organizers and launched campaigns around everything from poor roads and public education to welfare rights.[34]

Their unity and success, however, provoked a reaction. As the momentum of political change began to peak and politicians throughout the country lost control of volatile local constituencies, agents of a conservative backlash began to gather their forces. On the national level, Republicans in Congress persisted in attempting to dismantle OEO and parcel out its programs to preexisting federal agencies, a goal they would finally reach under the Nixon administration. Southern Democrats, whose support for the war on poverty had always been the product of complicated horse trading, began to balk at appropriating new funds for OEO. In 1967, as riots flashed through the ghettos of Detroit, Newark, and elsewhere, the legislative struggle over OEO reached a crisis. Amid allegations that they fostered "class hatred," provided leadership for the urban riots, and were infested with communists, independent community action programs became the sacrificial lamb that ensured renewed appropriations for OEO. Under the Green amendment (for Congresswoman Edith Green [D-Oreg.] ), nicknamed the "bosses and boll weevil" amendment, all funds for community action were to be channeled through a state or local government, or an agency designated by them. This appeased the multitude of politicians threatened by community action groups, brought several Dixiecrat votes back into the fold of the Democratic party, and eliminated any genuinely independent, federally assisted community action programs. The machine politicians had won.[35]

The influence of the war on poverty in Appalachia reached beyond its own constituents and short life span. Especially through community action and the volunteer programs, a collective, community-based model of political activism became legitimate and sanctioned:

> Community action was one of the important psychological preconditions for the black lung thing to happen. People in large

numbers were not working through conventional political mechanisms, working and making things happen that they never thought they could before.[36]

> Anybody who protested was a communist at least. It's the way to keep people down. What happened during the sixties, the right of protest became a respectable thing. I think by the time that the black lung issue became the number one thing, people were well versed about a lot of things—their rights. By the time of the black lung march up the Boulevard,* the unemployed fathers had camped on the statehouse grounds saying we want decent work to do, we want useful work to do, and we want to be paid for it.[37]

The war on poverty also popularized specific political beliefs that would appear again in the black lung movement. The notion that government has a social responsibility to the poor was reinforced. Aggressive, militant attitudes toward various benefit programs developed; scornful of bureaucratic procedures and complex eligibility standards, politicized poor people came to see government assistance as their right. Like many disabled miners and widows and, later, black lung activists, they invoked a history of sacrifice and privation to legitimate their demands for their "just due." Moreover, OEO's emphasis on the "maximum feasible participation" of the poor in running community action programs (the reality of which varied widely throughout the country) encouraged the notion of grass-roots, constituent control of federal programs. This principle became important in the black lung movement, during which activists repeatedly and sometimes successfully demanded changes in the administration of federal black lung compensation, and even participated in formulating regulations governing the program. Finally, one of the most solid links between the antipoverty programs and later political movements consisted of

* In 1969, during a wildcat strike for black lung compensation, miners, their families, and their supporters marched up the main thoroughfare in Charleston, West Virginia (Kanawha Boulevard), to the statehouse (see chapter 4).

VISTAs and Appalachian Volunteers who decided to make Appalachia their home, and who came to exert a critical influence over certain stages of the struggles for black lung compensation and union reform. First, however, it was necessary that the focus of political attention and activity in the coalfields shift to miners and their union, and this required the initiative of the miners themselves.

## The Growth of Insurgency, 1963–1967

John L. Lewis's retirement from the UMWA presidency in 1960 marked the end of an era in the union's history. During the decade that followed, protests against the UMWA leadership grew more strident and widespread, upsetting elections, conventions, and collective bargaining practices. For the first time in almost forty years, an insurgent candidate challenged the UMWA president for his office. Mounting rank-and-file concern over fund benefits, unemployment, and working conditions was aggravated by collective bargaining agreements that time and again were focused almost entirely on wage gains. Miners forced an end to Lewis's policy of flexible contract termination by wildcatting for new bargaining talks every two years; frequently they stopped production following—and in protest over—a new agreement. In sum, the complex factors that had created a bond of allegiance between the rank and file and their leaders—Lewis-conferred legitimacy, collective bargaining advances, patronage, and repression—were no longer sufficient to maintain loyalty and discipline, as miners directed at their new president a generation of accumulated grievances.

Tony Boyle became president of the United Mine Workers of America in January 1963, after Lewis's immediate successor, Thomas Kennedy, died in office. A product of UMWA District 27 in Montana, Boyle came to Washington, D.C., in 1948 to serve as a special assistant and, eventually, heir apparent to John L. Lewis. Boyle learned the role well, deliberately attempting to copy Lewis's ferocious rhetoric and demeanor. As president, he continued Lewis's traditions: collective bargaining contracts retained their exclusive focus on monetary gains, the internal structure of the union remained highly authoritarian, and nepotism continued to flourish, as Boyle appointed his brother Richard president of the nearly defunct District 27 and placed his daughter, a lawyer, on the district's payroll, where she eventually earned forty thousand dollars annually.[38]

Despite his efforts to imitate the master, Boyle began to have trouble only seven months after he assumed office. UMWA miners throughout the country worked under the 1958 wage agreement, essentially subsidizing the industry's bid for prosperity and larger market shares, until rebels in western Pennsylvania decided that five years under the same contract was long enough. Dissent originated in an area with a long tradition of support for autonomy, where coal mines and steel mills clustered together in one giant industrial nerve center. Workers could easily compare their respective wages, benefits, and working conditions, and by almost all counts, coal miners came up short.[39] By August 1963, miners in UMWA Districts 4 and 5 (western Pennsylvania) began meeting to plan a wildcat to force renegotiation of their contract. Wages were of secondary importance: "Unemployment comes first, and safety. The last thing we want is money. And that's probably the first thing they'll offer us."[40]

Intervention by the well-respected president of District 5, Joseph Yablonski, spared Boyle an embarrassing inauguration as UMWA president. Speaking to a meeting of about two hundred dissidents, Yablonski assured them that Boyle would give "serious consideration to [their] suggestions" and urged them not to strike.[41] Deciding to take his advice, a delegation of miners met with Boyle in September and received reassurance that he would reopen the contract. By mid-December, bargaining talks were underway, but rank-and-file pressure did not abate.[42] The content of the new contract continued to be debated at mass meetings of miners from Pennsylvania and northern West Virginia.[43]

Meanwhile, Boyle entered the 1964 contract talks amid great fanfare in the press, which predicted that he would do a "dramatic about-face"[44] for union policies by insisting that job security and safety take precedence over mechanization and wage demands. When he emerged with a new contract three months later, the UMWA president proclaimed that a new seniority clause provided that the worker "with the greatest seniority in the mine will be retained in case of layoff"[45] and that a new provision concerning helpers on certain machinery would generate employment. Both claims were exaggerated and misleading: the seniority clause stipulated that not only length of service but also "qualification to perform the work" might be considered in laying off workers, a loophole that left many management practices intact. The helper clause contained an explicit disclaimer concerning its impact on the size of the work force: "It is understood that the working crew is not to be increased by designation of a member as helper."[46] Furthermore, in the 1964 contract, for the first time the union formally affirmed its support for mechanization by giving blanket recognition to management's right to install any new machinery in the workplace.[47]

Discontented miners were quick to see through the union's deceptive assessment of the 1964 contract. Within a week after the agreement was signed, wildcat strikes began spreading east and south out of Ohio into Pennsylvania and West Virginia. At issue was not simply the absence of any forceful job protection provisions; miners also believed that Boyle had made a mockery of accountability by ignoring demands and priorities that they had carefully communicated to him. On April 1, fifteen hundred miners representing forty-eight local unions gathered in Bellaire, Ohio, and exhorted each other during a three-hour mass rally to continue the strike until the international officers came to the coalfields and explained the pact.[48] The strike pulled out eighteen thousand miners across six states but lost momentum when district and international officials coordinated a determined back-to-work campaign. Although miners failed in their contract objectives, this strike marked the beginning of a consistent escalation in wildcat activity that would continue for over a decade.[49]

Faced with an open rank-and-file rebellion and soon thereafter challenged for his office by Steve ("Cadillac") Kochis, a dissident from southwestern Pennsylvania, Boyle began retaliating with heavy-handed repression, for which he soon became known. Four months after the contract wildcats, delegates attending the union's quadrennial convention in Bal Harbour, Florida,[50] found Boyle prepared to stampede the opposition. The convention opened with an orchestrated tribute to the UMWA president that was worthy of the big top: forty-seven minutes of music from five bands (imported at a cost of $390,000) accompanied by a chorus of whistling, stomping, and calling from Boyle loyalists.[51] Packing the convention floor were delegates' wives, representatives from "bogey" (bogus) locals with only a handful of disabled or pensioned members, and approximately five hundred ushers, sergeants at arms, messengers, and committee members picked and paid for by the international.

Recognizing the paramount importance of the first day's proceedings, when convention rules and delegates' credentials were established, insurgents attempted to speak in protest against certain procedures. A few of those who managed to reach microphones found them guarded by marshals from District 19 (southeastern Kentucky and Tennessee) who were ready to silence dissent with blackjacks and fists. Others found that their microphones had simply been turned off. While marshals thrashed to the floor at least one miner, John Stofea from Pennsylvania, Boyle watched in silence from the rostrum, waited until the melee subsided, then presided over the rapid adoption by voice vote of the convention rules.[52]

This was the first and last significant attempt to challenge Boyle's authority at the 1964 convention. Substantive issues were hard to find during the remaining eight days of proceedings. Autonomy attracted little attention; those who spoke in its favor appeared daunted by Boyle's stranglehold on the union and the fact that their cause was more than thirty years old:

> DELEGATE MORRIS, DISTRICT 31 [northern West Virginia]: Mr. Chairman, I am going to say one word and sit down, because when you are beat there is no use going on. The record will probably show that my local union has asked for autonomy and we adopted it by official resolution. It is not my thought, for the delegates . . .—over five hundred of them—adopted this resolution by constitutional action after due notification, and they said they wanted autonomy. And I imagine they will want autonomy as long as there is a drop of American blood in their hearts. That is all I have to say.[53]

Resolutions concerning Welfare and Retirement Fund benefits far outnumbered those on any other subject and suggested a broad base of concern over cutbacks in pensions and medical coverage. However, most of those who were critical of fund policies simultaneously took care to affirm their trust in the international officers; Boyle had no difficulty referring their resolutions to the executive board and fund trustees and, hence, into oblivion. Similarly, despite opposition even from those who ordinarily supported him, Boyle engineered through the convention resolutions designed to solidify his power by lengthening his term of office from four to five years and upping from five to fifty the number of local unions required to nominate a candidate for international office.[54] He left the convention triumphant and went on to defeat Kochis by a vote of 96,000 to 19,000 in the December election.[55]

Despite these victories, the new year brought fresh evidence that the rank-and-file opposition would neither disappear nor be bludgeoned into silence. In September 1965, a rash of wildcats over various issues broke out in Pennsylvania, West Virginia, and Ohio. When six workers at the Hanna Coal Company's Ireland mine near Moundsville, West Virginia, were fired for participating in the illegal work stoppage, even more miners walked out. Within two weeks, more than fifteen thousand workers were on strike, and virtually all coal production in the area ceased.[56] Miners responded with scorn and anger to efforts by district and international officers to end the wildcat. An unidentified UMWA source in Washington, D.C., conceded to one reporter: "The strikes have gone beyond just a local grievance; it's a matter of the international's prestige and authority."[57]

Ironically, the business press seemed to perceive the significance of the rank-and-file turbulence much more clearly than the Boyle administration. In August 1964, *Business Week* was already predicting "The UMWA faces a miners' revolt,"[58] and by the September 1965 wildcats, the *Wall Street Journal* observed: "The soft coal fields are in the midst of a period of deep labor unrest that won't end without some sharp and fundamental changes in collective bargaining agreements and in the structure of the United Mine Workers Union."[59]

Boyle paid no heed. Controversy encircled the 1966 contract talks in what was largely a reenactment of the 1964 negotiations. Once again, dissident miners demanded large gains in job security, fringe benefits, and safety, placing little emphasis on wages. On Friday, April 8, 1966, Boyle made a surprise announcement that the union had concluded an agreement with three large midwestern producers and expected to obtain similar terms from the BCOA. A spokesman from the operators' association immediately rejected the pact as "too costly," and virtually overnight the largest coal strike in fourteen years swept across nine states. By Monday, April 11, more than fifty thousand miners had stopped work.[60]

The confused goals of the wildcat reflected the uneven

development of the insurgent movement: many miners simply wanted to assist Boyle in his effort to impose the midwestern agreement on the BCOA; some were striking in protest against the terms of the initial settlement; and still others were expressing their objection to the union's leadership, dictatorial structure, and overall collective bargaining policies. On April 23, Boyle concluded negotiations with the BCOA by accepting a pact that embodied the traditional priorities. A week of bitter debate ensued, during which an enraged minority of miners pushed for a continuation of the strike and district officials ordered the men to end it. By the first of May, most miners were trickling back to work.[61]

Over two years of relative quiet followed the 1966 contract controversy. No major strikes tested Boyle's authority as UMWA president, but the number of brief, localized wildcats continued to rise each year.[62] Beneath the placid surface, critical social and economic changes were shaping the future. Nourished by the Vietnam war, the growing utility market, and the UMWA's cooperation, the soft coal industry was beginning to prosper, and in the deep mines of Appalachia an ironic problem developed: a shortage of skilled labor.[63] The economic well-being tended to enhance miners' job security, make strikes a more potent weapon, and rankle workers already beginning to question the collective bargaining achievements of their union leaders.

This period also saw the beginning of a tremendous turnover in the coal mine labor force, as members of the generation that remained in the mines during and after World War II came to the end of their working lives.[64] Upheaval and change inscribed their collective biography. These men had seen the "mighty host" of 400,000 working UMWA members shrink to a mere 90,000. They had participated in massive, defiant strikes during and after World War II, only to find themselves disciplined for walkouts by their own union, which had refused to authorize a nationwide strike of any kind. In their contracts and at the workplace, this generation had witnessed the unwillingness of the UMWA to defend them against increasingly urgent problems like speedup, layoffs, and safety hazards. And now, as they faced retirement, these miners found that unemployment and migration had scattered the extended family and thereby destroyed the traditional protection against the worries of the aged. After decades of working in the mines and struggling to build and defend the UMWA, most miners had only a pension of $100 per month and a chest full of coal dust.

During this same period, disabled miners and widows in southern West Virginia began to agitate and to bring suit against the Welfare and Retirement Fund's pension policies. Working miners, especially in the heartland of discontent near the contiguous borders of Pennsylvania, West Virginia, and Ohio, continued to talk of reforming the UMWA's internal structure and collective bargaining goals. Furthermore, in 1968, three doctors began barnstorming the West Virginia hills to spread the word that the inhalation of coal dust causes black lung, a crippling, potentially fatal occupational disease. Into this tinderbox of agitation and converging social changes flew a spark—literally—that ignited the methane gas and coal dust in Consolidation Coal Company's No. 9 mine near Farmington, West Virginia. On a cool fall morning, November 20, 1968, seventy-eight miners perished beneath the earth in one of the worst mining disasters in the nation's history. The black lung movement was born.

# 4

## Resistance against Disease

On the night of November 19, 1968, ninety-nine men gathered in the dark near the portals of Consolidation Coal Company's No. 9 mine, ready to begin work on the "hoot owl" shift. Only twenty-one of them would return alive. At 5:30 the next morning, a violent explosion tore through the underground workings, igniting coal dust, methane gas, and the coal seam itself. The force of the blast shattered the main entrance to the mine—the Llewellyn portal—and fire and smoke spewed into the sky. Although twenty-one men escaped via portals remote from the explosion, hope for the remaining miners diminished over the next nine days, as the underground inferno exploded again and again. Even initial attempts to redirect the air flow and draw the fire away from the mine's main working area were unsuccessful: twenty-six tons of concrete and steel used to cap an air shaft were shattered and hurled into the air by repeated explosions.[1]

By the time a second major blast occurred, at 8:00 on the morning of November 20, friends and relatives of the trapped miners had begun their vigil at the company store across from the mine tipple. Close on their heels came medical personnel, news reporters, federal and state mining officials, politicians, coal company representatives, and union leaders. Television cameras arrived to record the grim activity. UMWA safety experts and engineers from Consolidation Coal, the U.S. Bureau of

Mines, and the West Virginia Department of Mines conferred over the best techniques to control the fires and explosions. Politicians and officials from the company and the union made brief, fatalistic speeches about the "inherent danger connected with mining coal."[2] Medical personnel dispensed sedatives to the grieving relatives.

News of the disaster at Farmington, West Virginia—the worst mine explosion since 1951, when 119 miners were blown up in a mine near West Frankfort, Illinois—was immediately broadcast across the country by newspaper, radio, and television.[3] For ten days, while attempts were made to control the raging fires and explosions, the hazards and sorrows of coal mining merited front-page and prime-time attention. This widespread media attention, along with the timing of the disaster, heightened the public impact. A nation of people celebrating Thanksgiving with family and friends found the latest football scores competing for news space with the hopeless reports from West Virginia. Reporters described and cameras panned the now ironic and pathetic-looking Christmas toys and decorations in the company store, which doubled as the disaster recovery center. In stories tinged with outrage, drama, and sensationalism, many journalists made the most of the "human interest" aspect of the disaster: "Baby Due Any Minute, She Won't Leave Mine."[4] Others were shocked by bland and self-serving comments on the explosion offered by government, company, and union representatives, and made sure to include in their stories the more incriminating examples of the official attitude.[5] The disaster motivated still others to investigate and report on the appalling history of the deaths and injuries in U.S. coal mines.[6]

UMWA president Tony Boyle arrived at Farmington with a mouthful of platitudes about Consol's leadership in mine safety and the unavoidable dangers of coal mining. Unlike John L. Lewis, who, regardless of his actual efforts to improve mine safety, was always shrewd enough to denounce "this abominable record of slaughter unequaled in the civilized world,"[7] Boyle seemed to defend the company, making the following statement at the scene of the disaster:

> This happens to be, in my judgment as President of the United Mine Workers of America, one of the better companies to work with as far as cooperation and safety is concerned. This was proven when they aided the UMWA in Congress by supporting amendments to strengthen the Federal Coal Mine Safety Act when numerous other coal companies were fighting our bill.* It's a sad commentary that this thing had to happen anywhere, but more especially here. . . . As long as we mine coal, there is always this inherent danger connected with mining coal.[8]

Boyle's blasé speech at the mine site drew the hot spotlight of press attention. Joseph Loftus, writing in the *New York Times*, observed: "The United Mine Workers has not made a significant fight for safety in some years. It has virtually adopted the position of the operators that safety improvements that cost money would hurt coal's competitive position and cause unemployment."[9] Public officials such as Congressman Ken Hechler (D-W.Va.) and Senator Gaylord Nelson (D-Wis.) also criticized the UMWA's record on health and safety, to which Boyle replied with countercharges of political opportunism:

> A few national figures took advantage of an opportunity for publicity to take a "cheap shot" at the Union, management, and Federal and state governments. While those four groups were working feverishly together in an attempt to rescue the trapped men[,] the "instant experts" were being interviewed and filmed in Washington and New York far from danger and far from any knowledge of coal mining.[10]

While this war of words got underway in the press, friends

* These amendments to the 1952 Mine Safety Act extended its coverage to small mines. It was, of course, in the economic interest of Consol and other large companies to lobby for the bill (see chapter 2).

and relatives of the Farmington miners continued their vigil. Air samples drawn from deep inside the mine revealed lethal concentrations of methane and carbon monoxide. Sensitive listening devices inserted through drill holes picked up no sounds of life. Despite thirteen violent underground explosions in the space of five days, rescue teams ventured briefly into the mine, but they found no trace of the trapped miners. Pleas from the miners' relatives to continue the rescue efforts were to no avail. On November 29, crews began to close off the mine in order to extinguish the underground fires and to prevent further explosions. Seventy-eight men were thus sealed in the underground tomb of the Farmington mine.

## Famous Names and Forgotten People: The Emergence of Black Lung as an Issue

The Farmington mine disaster has gone down in history as the catalyst for the most far-reaching coal mine health and safety legislation ever passed in the United States. Many analysts also view the disaster as the "precipitating event" of the black lung movement.[11] The explosion drew prolonged media attention to the problems of coal mine health and safety and helped to make black lung disease a national issue. It galvanized the forces of reform within the U.S. Congress and thereby contributed to the eventual success of federal health and safety legislation. The major impact of the Farmington disaster seems to have been at the national level; among those in the southern West Virginia coalfields who became active in the black lung movement, few attributed their initial involvement to the explosion or even mentioned the explosion in their oral histories.[12] Indeed, reporters covering the Farmington mine disaster were able to pick up on the problem of black lung because miners and a few supporters had already begun to make it an issue.

"Asthma," "silicosis," "rock dust"—by whatever name, coal

miners had long recognized black lung as a disabling, sometimes fatal, occupational disease associated with dust inhalation (see chapter 1). The black lung movement of 1968–69, although more determined and successful than previous efforts, was not the first time miners had collectively tried to combat the problem. They first sought to prevent black lung during the late nineteenth century, through legislative regulations on underground ventilation systems. As the United Mine Workers of America developed into an effective base of power, miners' strategies shifted toward industrywide unionization, contractual requirements concerning ventilation, and, in some cases, nationalization of the mines. Mechanization and the increased division of labor aggravated the dust problem and provoked additional responses from miners: sabotage; refusal to grant machine differentials; and demands for removal of machinery from the mines (see chapter 2). Unable to stop mechanization, miners ceased their active opposition to machinery per se but continued to push for reduction of the dust. For example, in 1948, miners from Terre Haute, Indiana, made a farsighted plea for mandatory dust suppression to the UMWA convention:

> WHEREAS, Owing to the hazardous condition of mining affecting health, such as foul air and dust, which cause asthma and silicosis poisoning, therefore forcing the miner to give up his occupation, being disabled through no fault of his own; therefore be it
>
> *Resolved,* That this convention go on record as favoring the passage of a national and state law to compel mine Operators to install equipment to eliminate dust on cutting and loading machines.[13]

By the late 1930s, contention arose concerning workers' compensation for black lung disease. The widely publicized tragedy at Hawks Nest Mountain, West Virginia, where an estimated two thousand workers were killed or disabled from silicosis, generated numerous damage claims and calls for coverage of this disease under state workers' compensation.[14] Believing themselves disabled by a work-related respiratory problem that many called silicosis, miners began filing claims under the new compensation provisions that were enacted by most states during the latter half of the decade. Almost everyone who filed, however, was turned down. The restrictive eligibility criteria applied to respiratory disease victims under the workers' compensation system created an undercurrent of discontent that would persist for the next three decades. From 1942 to 1968, coal miners regularly demanded at UMWA conventions that black lung disease be recognized, prevented, and made compensable.[15]

Until the late 1960s, most coal-producing states continued to award occupational lung disease compensation only for silicosis.[16] Forty-year veterans of the mines who were so disabled that they could not walk up stairs or sleep in a prone position were denied compensation because their X-rays did not reveal the classic pathological changes associated with this specific disease. Even the disabled who showed evidence of silicosis rarely qualified for a lifetime award based on total and permanent disability. Many applied some time after they had retired from the mines, when their lung disease progressed to the point of causing severe debilitation; they were turned down because statutes of limitations typically restricted the time period between the last occupational dust exposure and the filing of a compensation claim. Most others received only a partial disability award—in West Virginia, $1,000 for first-stage silicosis, $2,000 for second-stage. Unable to live on such a sum, they continued to work in the mines until some became so incapacitated by black lung that they had no choice but to quit. In many states, workers in this situation were prohibited from reopening their claims for compensation; a lump sum for partial disability was all they ever received. This restriction inspired some coal companies to require that experienced job applicants file silicosis claims against their previous employers; if the claims were approved, the company need not fear that it would incur liability, because miners could make no further claim.[17]

The injustices and inequities of this situation were magnified when the generation of miners who worked in the dusty, mechanizing mines of the postwar period began to retire. There is little doubt that the prevalence of occupational lung disease among this generation was higher than among most others in the past.[18] Their common experience of unimpeded mechanization and uncontrolled dust became their shared hardship of physical disability and denied compensation. Some coalfield physicians even continued to regard "miners' asthma" as a nondisabling condition and informed their patients that coal dust was harmless, if not beneficial. It is highly unlikely that they performed the diagnostic procedures or provided the medical documentation necessary for miners even to pursue compensation.

During the same period, however, the research of a few dissenting physicians began to confirm the longstanding experiential knowledge of those who had lived with and died from black lung disease. In 1964, local newspapers in southern West Virginia first reported on the U.S. Public Health Service's study of CWP among Pennsylvania miners and on the research findings of certain fund-affiliated physicians: scientists had "discovered" a "new" occupational lung disease among coal miners.[19] The emotional impact of these reports was profound. For the disabled who had been denied compensation, the news lent a righteous legitimacy to their anger. For those accustomed to humiliating diagnoses of "malingering" or "fear of the mines," "they was good words to hear, believe it or not, because nobody had been saying it."[20] Others simply felt confirmed in their self-diagnosis:

> It's just like this: all down through the years, we'd been told that dust has no effect. They'd say you cough it up; it won't stay on your lungs. I moved here in the forties, and had to walk up this hill to catch a ride. I was working the night shift. Well, gradually I noticed I

> was losing my breath, couldn't hardly walk up the hill.
>
> Then, in the early sixties, the government appropriated money and made a study. They came up with this here headline, that miners had a disease, that they were going to die of it. I was convinced without any doctor's authority, as soon as I saw this article, that I had black lung.[21]

In 1968, the year that brought to a violent climax nearly a decade of social upheaval in the United States and many other countries around the world, coal miners and their families began to act collectively on the black lung problem. Once again, they turned to the United Mine Workers of America, their greatest source of institutional power, and requested that the union make black lung compensation and prevention a priority. At the 1968 UMWA convention, delegates from Shawneetown, Illinois, to Keen Mountain, Virginia, submitted eighteen separate resolutions concerning black lung. Twelve demanded disability compensation and put forward various strategies to achieve it: reform of state workers' compensation laws; increase in the operators' royalty contributions to the Welfare and Retirement Fund; and direct payments from coal companies to disabled workers. The six other resolutions addressed prevention of disease in the workplace. They called for contractual changes like mandatory dust suppression, establishment of a dust standard, and assignment of as many miners as necessary to the task of dust control.[22]

The UMWA leaders' response to the rank-and-file sentiment concerning black lung influenced the goals and especially the strategy of the movement that developed thereafter. Unable to nudge their union into action, miners were forced to develop leadership and organization on their own. President Boyle apparently anticipated their discontent at the 1968 convention and attempted to steer it toward what he thought would be the safe harbor of state-by-state legislative reform. Following a stirring speech on coal workers' pneumoconiosis by Dr. Lorin Kerr of the Welfare and Retirement Fund, Boyle called on the UMWA safety director to read a prepared resolution. It urged all districts to campaign for passage of model legislation that would establish black lung as a compensable disease under the workers' compensation laws of every coal-producing state.[23] Omitted were any calls for actual changes in the workplace, such as dust suppression to prevent black lung, which would have upset the union's established policy of industrial collaboration. The resolution passed unanimously; its stated goal, subsequently ignored by union officials, would become that of the rank-and-file movement over black lung.

UMWA leaders' failure to follow through on the delegates' decisions forced miners to take action on their own: "See, the miners themselves were the ones really agitating for it. They came back from the Denver convention all fired up. They thought the union was going to lead the way."[24] The union, however, did not lead the way. According to one miner who went with others to press UMWA District 17 to pursue compensation reform, one official* threw them out with the words: "I ain't sticking my neck out for you fellows. Get out of here!"[25]

Joe Malay, Ray Stall, and other miners from Fayette County, West Virginia, now went outside the union for help. They turned first to Isadore E. Buff and Donald Rasmussen, local physicians who were known to take seriously miners' belief in a widespread occupational lung disease. Working through the honeycomb of UMWA locals in southern West Virginia, Malay and others set up meetings of miners at which they invited the doctors to speak about black lung. The initial response was not encouraging:

> I wrote to about twenty locals, and I got one local to come. That was old Woody Mullins', down here at Gallagher. He listened to Dr.

* All district officials in West Virginia were appointed by the UMWA president; they had little accountability to the rank and file.

> Buff and said, "Would you come to my local and explain it?" "Yes, you just name the time." We told Dr. Buff we'd try to get a better audience. So we invited Dr. Rasmussen and a few other locals. Had more the next time, but still 25 would cover it.[26]

In the coal camps of adjacent Raleigh County, black lung was a topic of discussion at the meetings of the Association of Disabled Miners and Widows. Alerted by these discussions, VISTA worker Craig Robinson and lawyer Rick Bank began to talk about legislative reforms that would remedy the problems of unrecognized occupational lung disease and denied compensation. Aware of the scientific conflict over what constituted black lung, Robinson and Bank attempted to formulate legal presumptions that would give those seeking compensation the benefit of medical doubt. Robinson soon drew miners and a few other VISTA workers into the discussions; he called a meeting at the Mabscot Community Action Office, where Dr. Rasmussen spoke. As one activist recalled, "Not too long after that [first meeting], a smaller group sat down with Rick, began working out the basic concepts we had to have in the law—the language, the presumptions. . . . We were building toward introducing something in the legislature."[27]

Meetings remained small and scattered until late December, when there was a turning point in the black lung movement. At a meeting in Marmet, a little town on the Kanawha River near Charleston, the miners' political mood began to shift. One of the organizers, Woodrow Mullins, a coal miner who has since died from black lung, described this meeting:

> I got the community building in Marmet. They charged $50, but they let us have it for free. My daughter wrote letters to most of the senators and members of the House of Delegates. It was a pretty good crowd. And we sent posters to the local unions, as many as we could, and invited them. Dr. Rasmussen brought some lungs, real lungs with him, and from then on the meetings started picking up. They commenced believing us.[28]

The turnout and the spirit at Marmet gave the movement's organizers faith that miners were ready to move. Dr. Buff encouraged the miners to take action: "This movement must come from the miners themselves. The union and companies have had their chances and have done nothing."[29] Advised by Buff to hire a lawyer who could translate their discontent into a specific legislative agenda, and unaware of the prior work of those in Raleigh County, a small group of miners met at the City Hall in Montgomery to discuss how to raise the necessary money:

> The meeting just about broke up and everybody started home. Some were afraid we just couldn't raise the money. So we got down on the street—Charles Brooks, Woody Mullins, Lyman Calhoun, and a few others. We stopped on the corner and we said we just couldn't be defeated. If we had to sell peanuts on the corner, we had to raise that money.[30]

They succeeded. The group decided to found a new fund-raising organization and call it the Black Lung Association (BLA). They elected Charles Brooks, a black miner who was president of the Winifrede local in Kanawha County, as president; Ray Stall became vice-president, Ernest Riddle, treasurer, and Raymond Wright, secretary. All of these men were miners from Kanawha and Fayette counties. To raise money for the initial payment to the association's lawyer, Charles Brooks mortgaged his own house. The remainder of the $10,000 fee they raised by contacting miners in local unions up and down the Kanawha valley. Arnold Miller, president of the Bethlehem Steel local at Kayford, persuaded his members to contribute $1,000, and other locals came up with similar amounts.[31]

The movement swept through the coal camps of southern West Virginia in a matter of days. On the first weekend in January 1969, 450 miners rallied at the community center in Vivian; they decided to regroup in Charleston and march on the state capitol to demand a black lung law. Similar plans began to take shape on the same day at meetings in Delbarton and Logan. Every weekend thereafter they gathered: 200 in Chelyan, 450 in Madison, 750 in Pineville . . .[32]

> The turnout was absolutely unbelievable. Meetings used to be huge. For example, down at the community center in Vivian, that place was just jammed and overflowing. Buff got the national news media there. And then one weekend, I remember a meeting at Pineville. People were packed in, out in the hall, all over the place.[33]

Featured attraction was a "real dog and pony show," put on by the doctors. Buff, a born showman, transported around a suitcase of theatrical paraphernalia, including a set of lungs and a pair of hats—a black hat for when he pretended to be a politician talking to the "cold operators," and a white one for when the politician spoke to miners:

> We would open with a prayer. Buff would take them lungs and act . . . well, he went a bit too far sometimes. It's like this, you see, there wasn't but a very few people ever saw a set of lungs. He told the men about the bad lungs and he'd yell, "Feel 'em! Feel 'em!"[34]

Word of each meeting circulated not only through the leaflets and posters of its organizers, but also through the social bloodstream of the area. From the first small group of participants, the black lung movement radiated out through the crisscrossing social ties of the rural coalfields:

> Mr. Calhoun's wife was a cousin to my husband. And Jeep Hall worked with Mr. Calhoun at Burnwell, up on the creek. Mr. Mullins, he lived over in middle Gallagher, next to Mr. Calhoun. And Lonnie Sturgill lived at Montgomery. Mrs. Mullins knew Lonnie's wife that used to live in Gallagher. And Mr. Mullins was in the same local as Charles Brooks.[35]

Women soon joined in the movement, begining with the wives of the initial organizers:

> In 1969, women wasn't going to as many things as they are now. So Mr. Calhoun used to report in to me, call me and tell me all about what took place. One day he said, "Why don't you come to the meeting yourself?" I said, "I can't go up there. That's just a bunch of men. No women comes." And he said, "Well, Mrs. Malay comes." So his wife and Mr. Mullins' wife and Mr. Sturgill's wife and me all started going.[36]

The physicians, now three in number with the addition of pathologist Hawey Wells, soon became identified in the press as the leaders of the movement. This was not an accurate reflection of behind-the-scenes initiative and decision making, which continued to be the domain of many of the initial activists and organizers. Nevertheless, the physicians, and especially Dr. Buff, played a unifying role as public symbols and, sometimes, spokespersons for the movement. At a time when UMWA district officials were already accusing the Black Lung Association of dual unionism, visible leadership by persons with no union affiliation was a politic, if not deliberate, arrangement.[37] Even more important, the physicians played a central ideological role: repeatedly, in rally after rally, they conferred the legitimacy of their professional knowledge on miners' own experiences of black lung. This underscored the injustice of widespread suffering from an unrecognized, uncompensable occupational disease and made the legacy of company doctors appear more reprehensible than ever before.

Black lung activists from all over the state came together for the first time on January 26 at the Charleston Civic Center.[38] A crowd estimated at between twenty-five hundred and five thousand listened as miners sang songs about black lung, physicians described and deplored it, and a few maverick politicians threw in their lot with the movement. Now they were ready to act. The rallies had made black lung the talk of virtually every household in southern West Virginia. They had forged the individual experiences of illness, disability, and poverty into collective anger over injustice. They had unified participants around belief in their common entitlement to redress, and they had presented black lung compensation as the form it should take.

> Well, I probably was just fighting mad. Because, you know what got me. . . . You go to these rallies, there'd be these old men there, much older than my husband, couldn't get no breath at all. You'd think they'd go through the ceiling trying to get air. Before they'd get done telling you how dirty they'd been done, they'd be crying. Couldn't work; didn't get no pension; didn't have nothing to live on. Now, that gets next to you. It still makes me mad when I think about it![39]

## Climax: The Black Lung Strike

By mid-February, the West Virginia legislative session was more than half over, and there had been no apparent progress toward black lung reform. Numerous individuals and organizations, including the Black Lung Association and the United Mine Workers of America, had initiated separate bills concerning black lung compensation, but all of these bills were stuck in the House Judiciary Committee. Well aware of the forces committed to defeating their cause, miners became increasingly restless as the days dragged by and neither the House of Delegates nor the state senate passed a black lung bill of any kind. Talk of a wildcat strike began to circulate through the mines. On the day legislative hearings on black lung were held, the entire day shift at two mines stayed away from work to attend the proceedings. Several miners appeared at the capitol in Charleston with placards threatening, "No Law, No Work."[40] Three days later, about fifty miners marched on the capitol and warned that there would be a national shutdown of the coal industry if the legislature failed to pass the miners' black lung bill.[41] Still, legislation languished in the House Judiciary Committee, chaired by the powerful and conservative delegate J. E. ("Ned") Watson, the great-grandson of the founder of Consolidation Coal Company.[42]

Finally, on Tuesday morning, February 18, 1969, a local dispute at the East Gulf mine in Raleigh County angered 282 miners into striking. Reporters attempting to cover the event received conflicting accounts of its cause. One miner was quoted in the morning paper: "We feel like we should support Dr. Buff with some action. The legislature is not bearing down, they're letting it cool off too much."[43]

In a chain reaction, within hours, mines began closing at Itmann, Eccles, Slab Fork, and other adjacent communities in Raleigh and Fayette counties. By Tuesday night, nine mines were down. On Wednesday, roving pickets, word of mouth, and brief accounts in the news helped spread the strike west and south into Wyoming, Logan, Mingo, and McDowell counties. Within three days, ten thousand miners were on strike.

> It spread like wildfire after it hit the news. I think mainly what really ignited it was the fact that we had organized and come out publicly and the men at the mines saw the goddamn coal operators couldn't touch us, then they really come out. It just spread like wildfire.[44]

Behind the seemingly spontaneous mass action were the efforts of countless miners who organized themselves into roving pickets. The southern mines shut down in a matter of days, and miners headed north to other coal-producing counties.

> We run two brand new cars out working on this thing. We went up to Morgantown to get those boys to help us. They didn't cotton to us too well at first, but then when the strike got going, they helped us out.
>
> At that time, this was the biggest coal state in the country. If nothing else, we'd go up there, go to the mine and get with one man,

> whether we knew him or not, and explain it to him. And he'd say, "Okay, you boys go back to Mingo County or wherever you belong and we'll take care of it here." The next day all the mines over there would be out.[45]

At the beginning of the walkout, there was a significant but abortive attempt to broaden the movement to include other workers and their occupational diseases by pulling a general strike in the mines, foundries, and chemical plants that surround the state capital of Charleston. This effort reflects the heady atmosphere of political potency and solidarity among workers that infused the black lung movement in these days.

> I don't know who started it, but somebody mentioned, let's go shut the Alloy plant down, because they had mines too. So they wanted to go and shut the plant down. We was going to get it for occupational disease, any lung disease. Anyone that had anything wrong with their lungs would get it.
>
> What we intended to do was to shut the whole place down. All these places. Our idea was if we could shut Alloy down, we could shut the whole valley down—Belle, South Charleston, all of it. The idea was to make it a general occupational disease, you know. But it didn't work. They just weren't about to come out.[46]

Six days into the strike, on Sunday, February 23, twelve hundred miners—many of whom were local union officers and all told represented more than twelve thousand men—gathered at the union hall in Affinity, Raleigh County. Five hundred miners jammed themselves into the small frame building while seven hundred waited in the snow outside. The speakers, who included the three doctors, the officers of the Black Lung Association, and Delegates Ted Stacy and Warren McGraw, were forced to climb in and out through a window to take their message to both crowds. The spirit of camaraderie and militancy intensified with each speech. Dr. Buff roared at the crowd: "It's up to you men to stand on your own feet and fight, fight, fight, *fight*! The legislature hasn't changed since 1870. They don't really care. But they'd better care right now because you're going to get it. You've got to get it! Slave or free?"[47]

The Black Lung Association officers spoke more personally, about their experiences as coal miners, as black lung victims, as union members. Ernest Riddle, treasurer of the association, preached to the crowd:

> We appreciate every bit of support, we appreciate everything we see here today. And we know that everybody that's got a father—each and every one of you—when Dad comes home in the evening and he can't breathe, you know what's a-hurtin' him ["Yes, sir," "That's right"]. And a woman that's living with a man and seeing him slowly die, nobody has to tell her. Nobody has to explain to them babies what was wrong with Daddy that he couldn't get his breath at night. . . .
>
> Men, to make a long story short, I appreciate seeing all of these men out here today in support and expressing your appreciation for these doctors. I've only got an eighth-grade education. I ain't got much to lose, buddy. They done wore out everything that I have. These coal companies done wore out everything I have, and what they ain't wore out they plugged up. Now, I think there is a time and a place for everything, as the Bible says, and I think we have come to the crossroads where we must take a stand and put up a fight for what is ours, what is justly ours. I thank every one of you.[48]

Although the meeting was somewhat chaotic, as it lacked an agenda, clear leadership, and formal lines of authority or responsibility, there gradually emerged a plan of action: the miners would meet again on Wednesday, February 26, at the Municipal Auditorium in Charleston, march on the capitol, and, by their presence, pressure the legislators into acting on their black lung bill. Meanwhile, roving pickets would spread the strike by keeping out the midnight shift on Sunday at major mines in northern West Virginia. Those who could would come to the statehouse on Monday and maintain the pressure on the legislators.

The statehouse in Charleston became the battleground, as opponents and supporters of black lung reform descended on the legislature. Opposition to the movement originated largely from three predictable sources: the United Mine Workers of America hierarchy, the coal industry, and the medical profession. In all three cases, the popular uprising over black lung was objectionable not only because of its substantive goal, but also because it threatened the foundations of institutional power.

For the hierarchy of the union, the coal miners' aggressive action around black lung implicitly challenged their legitimacy and autocratic power as leaders, and threatened to disrupt two decades of stable relations between the union and the coal operators. UMWA officials first attempted to undercut the movement by drafting a separate bill for the legislature and rallying miners behind it. They then declared open season on black lung activists by charging them with dual unionism and forbidding locals to donate money to the Black Lung Association.[49] All of these efforts failed to weaken the momentum of the rank and file. The union leaders' greatest impact was probably to encourage disaffection among miners in southern West Virginia, and thereby prepare the way for union reform.

For the coal operators, the wildcat strike was obviously troublesome: not only did it shut off coal production, it indicated that miners were restive and possibly ready to walk out over issues other than black lung. Soon after the strike began, the operators filed suit in federal district court seeking damages of $1.1 million per day from the UMWA International and Districts 17 and 29.[50] Concern over the immediate impact of the strike was probably overshadowed, however, by the operators' long-range interest in defeating workers' compensation for a broadly defined black lung disease, a potentially enormous financial liability. Assuming the mantle of scientific caution and rationality, the West Virginia Coal Association declared:

> The intelligence of all West Virginians—particularly that of the legislators—has been insulted by such things as nationwide TV appearances which said, in essence, that West Virginia's legislators are two-faced; the miners' so-called "March on the Capitol"; . . . a mock funeral and the placing of a coffin in the rotunda of the Capitol; and the flamboyant displaying of so-called diseased lungs to the press and legislators.
>
> . . . [The] other side—the vast majority of physicians—has tried to to present scientific, medical facts relating to coal workers' pneumoconiosis. . . . The people of West Virginia . . . should try to pass just legislation which is based on the solid medical facts of coal workers' pneumoconiosis, not sensationalism and emotionalism.[51]

The black lung movement posed its greatest ideological threat to the medical profession, which proved a formidable opponent. At issue was not only scientific disagreement over the nature and extent of occupational lung disease among coal miners. This populist, working-class uprising threatened to intrude upon the inner sanctum of physicians' status and authority—their professional control over the definition of disease. To the medical establishment, the trio of physicians who urged the movement on were especially abhorrent. They violated the rules of scientific discourse by using political force to establish the existence of a disputed disease. They threatened the scope of medical practice by supporting legislation that would reduce a physician's diagnostic role in assessing occupational disability. They undermined professional credibility by continuously excoriating "company doctors." And they challenged the profession's hard-won control of legitimate medical knowledge by popularizing an unscientific disease concept called "black lung" and encouraging miners to believe in it. For a profession whose close ranks have long stood out as unique in the United States, the physicians' hostile condemnation of their fellow practitioners Wells, Rasmussen, and Buff was remarkable.

The Cabell and Kanawha county medical societies both passed resolutions condemning the "alarming appeals based upon evidence open to serious question of its validity." Urging the Cabell County Medical Society to adopt this resolution was Dr. Rowland Burns, who stated: "There is no epidemic of devastating, killing and disabling man-made plague among coal workers. . . . It is my opinion that the false prophets and deluded men who present their hypotheses concerning technical medical problems to the general public without discussion or presentation should be condemned."[52] The Kanawha County group formally notified the West Virginia legislature that they opposed coverage of coal workers' pneumoconiosis under the state's workers' compensation laws.[53]

All of these opponents to the black lung movement came out in force at the state capitol. It was the miners' presence, however, that was overwhelming. One day after the meeting at Affinity, hundreds of coal miners descended on Charleston, roamed through the halls of the statehouse, and packed the galleries in the House of Delegates. A legislator who supported the black lung movement described the atmosphere:

> The miners started flooding into Charleston, lining the galleries in large numbers. You know what happens when you get large numbers of people together and they're discontented. There was a lot of noise, a lot of talking. It was like sitting on a razor's edge. These guys, the miners, had been waiting in the galleries and they didn't know what was going on. They didn't understand about first readings and second readings and third readings and all this.
>
> Meanwhile, the big oaken doors had been shut, pulled on rollers. This wasn't like just closing any door, you had to roll these heavy things shut. And the State Police were in the cloakroom in the House of Delegates, all because the miners were there. You have to understand the situation in there. It was really getting bad. I mean, if

> somebody had stood up and said something really inflammatory, there would've been a riot.
>
> So I stood up and told the Speaker, "I see the doors are closed and there's an armed guard in the cloakroom." I said, "Open the door so I can go out 'cause if there's going to be trouble in here I want to be out there with my friends." Well, that brought down the house.[54]

On Wednesday morning, February 26, convoys of buses and cars carrying miners from all over the state began arriving in Charleston for a rally. By this time, the strike was one week old; coal production in southern West Virginia was at a standstill, and over half of the northern mines were shut down. More than two thousand miners gathered at the Municipal Auditorium that afternoon to hear the familiar round of speakers. Carrying placards reading "No Law, No Work" and "78 / 4" (representing the miners killed at the recent Farmington and Hominy Falls disasters), miners poured out of the Municipal Auditorium and marched down Kanawha Boulevard, the main thoroughfare in Charleston, to the state capitol. As they passed the headquarters of UMWA District 17, they booed and taunted. Even as the miners marched, members of the House Judiciary Subcommittee were rushing to the floor a black lung bill. One legislator recalled viewing the march from the statehouse: "We were in a committee meeting. Our room was on the third floor facing Kanawha Boulevard. You could look out and watch them come—watch them march up the steps and into the building. . . . Well, none of us had ever seen anything like that!"[55]

The explosive atmosphere in the statehouse brought West Virginia's Republican governor, Arch Moore, to the steps of the building to greet the miners. His efforts to reassure them by stating that he would introduce a black lung bill during a special session of the legislature if the current bills failed to pass brought forth a unanimous roar: "No! No! No! Now! Now! Now!"[56] The miners swept past the governor and up to the locked doors of the chamber of the House of Delegates:

> We marched up there and there was State Police at the doors. But with that many of 'em the State Police just stepped aside. They had the doors locked at the House of Delegates. The men just walked up there and asked the guards, they said, "This is a public place. Are you going to let us in, or do you want us to kick it down?" They let us in.[57]

The miners' action broke the political logjam in the House Judiciary Committee, which reported out a black lung bill that same day. Elation at their victory soon gave way to indignation, however, when the bill's content became clear. It offered compensation only to victims of pneumoconiosis, not other potentially work-related lung diseases, and it gutted the presumption clause, which the BLA's legal advisors considered essential. The miners resolved to push for amendments on the floor of the house, and they issued a formal statement to the press:

> By mistake, mischief, or malice, the House of Delegates Judiciary Committee has recommended a workmen's compensation bill which does justice only to the West Virginia Coal Association. This atrocity is an insult to the miners who have worked hard and sacrificed much to obtain decent legislation. If passed, the committee bill would leave the miners worse off than it found them.[58]

By Friday, when the bill was due for a vote in the house, coal production had ceased throughout West Virginia, and over forty thousand workers were on strike. At the statehouse, miners ringed the galleries and exerted the pressure of their presence on members of the House of Delegates as they considered amendments to the bill. The legislators finally gave ground on several points, and an amended bill similar to the one originally proposed by the Black Lung Association passed the house by a vote of 95 to 0.[59]

Now it was the operators' turn to pounce on the legislation. Soon after the beginning of the strike, the West Virginia Coal

Association had made it known that its members would accept black lung reform, but they had also enlisted a platoon of lobbyists to defeat the progressive language of certain bills. The legislation that passed the house contained almost every clause they opposed. The Coal Association denounced the house action as "a classic example of bowing to pressure. . . . [It is] a dire warning to all businesses and industries in the state. It has opened the Pandora's box." Association president Quinn Morton stated that the bill was "founded on sensationalism and irresponsibility, nurtured on emotionalism and passed under a gallery of pressure with total disregard of proven medical facts." It was, in short, "galloping socialism in one of its purest forms."[60]

At stake in this political tug of war was the scope of eligibility for black lung compensation.[61] Movement activists sought legislation that would rectify decades of legal omission and circumvent persistent medical conservatism by identifying as compensable a broad "coal workers' lung disease." No specified diagnostic test, such as a positive chest X-ray, would be required to establish eligibility. This legal approach conformed with the medical perspective of doctors who argued that "black lung," that is, occupationally related lung disease among coal miners, included more than the single disease of CWP and required diagnostic techniques other than X-rays. The BLA also proposed a legal presumption that would assist claimants in establishing that their disability was work related (which was an eligibility requirement for workers' compensation). In the most liberal bill, the lung disease of miners who had two or more years of continuous occupational exposure to coal or silica dust and evidence of a "coal workers' respiratory disease" was presumed to be related to their work. Chronic lung disease can arise from many sources in the workplace, home, and elsewhere, which often makes it extremely difficult legally to prove occupational origins. The presumption gave miners the benefit of the doubt to make up for medicine's inability to pinpoint a single cause or apportion blame among several causes in each individual case of lung disease. They also sought to liberalize the statute of limitations that required silicosis claimants to file within one year of the date of their last occupational exposure. In the context of a hitherto unrecognized and uncompensable disease, the statute created an impossible situation for retired miners: those disabled from black lung who had never received compensation because it did not exist would now be ineligible because they were too late to file a claim. All of these legislative proposals reflected the movement's belief that coal operators were ultimately to blame for black lung disease and should therefore compensate all who were disabled; conservative medical opinion, which had been part of the problem from the beginning, should not be allowed to stand in the way.

Coal operators, whose financial interests made their view of black lung coincide with the medical profession's skeptical and restrictive view of it, pushed for legislation that would recognize at most coal workers' pneumoconiosis as a compensable disease. Invoking the ideological authority of physicians who argued that "cigarette smoking is the most important factor" in the development of miners' respiratory disease, the operators fought any presumption of work relatedness. Also on their side was legal precedent, in the form of the statute of limitations; the operators, of course, contested activists' efforts to lengthen the period for filing a claim, which could potentially open up compensation to every aged coal miner in the state of West Virginia.

Legislators conducted their eleventh-hour bargaining under extreme pressure from lobbyists on all sides. Hundreds of miners paced the halls of the statehouse and filled the galleries. In addition to their customary lobbyist, the locally well known firm of Jackson, Kelly, Holt, and O'Farrell, member companies in the West Virginia Coal Association sent their own executives to the capitol to thwart any far-reaching black lung reform. One legislator said of the effort: "We were really bombarded by lobbyists. I never knew the industry had so many. They were just all over the place."[62]

As four hundred miners looked on from the galleries, the senate passed a relatively weak black lung bill by a vote of 34 to 0 on Wednesday, March 5. This bill, which emphasized X-ray evidence of pneumoconiosis and eliminated the presumption clause, seriously eroded the BLA's goals as embodied in the house bill. By Friday, a conference committee made up of members from both legislative houses was still unable to reach agreement on a compromise bill. One day remained in the session. The Black Lung Association stepped up its pressure by warning that the strike would continue indefinitely if an acceptable bill failed to pass. Activists also sent a telegram to President Richard Nixon, asking him to declare the entire state a "disaster area" because of unsafe conditions in the mines, and petitioned Boyle to call a national work stoppage in commemoration of the seventy-eight miners killed at Farmington.[63]

At ten minutes before midnight on Saturday, when the legislative session was due to adjourn, the conference committee rushed a bill to the floors of the West Virginia house and senate. This bill extended compensation to miners disabled from "occupational pneumoconiosis," included a weak presumption clause concerning the work relatedness of disease, and extended the statute of limitations to three years.[64] The legislation passed with only two votes in opposition.[65]

Miners and other activists now had to decide whether to accept this bill or continue the strike. On Sunday, March 8, more than two thousand people crowded into the Park Junior High School in Beckley for what would be their final rally.

> The biggest, the most stirring meeting that I remember was on the night after the end of the legislative session at the junior high school. I've never seen a gathering like that—not only was it just packed, but, oh, the spirit! There was absolutely no standing room. The

whole auditorium, the basketball court-sized stage behind the auditorium, all of the hallways, and even out onto the sidewalks were filled with miners. I remember there was one guy that got up and asked for a vote to go back to work. And I can't describe it . . . but there was a great big roar—"No!"—that they wouldn't go back to work. They were going to wait until the governor signed the bill.[66]

The leaders of the Black Lung Association and the few professionals who had served as speakers at the rallies and advisors to the association were in an extremely difficult and pivotal position at this point. Miners looked to them for an analysis of the legislation and for guidance on continuing the strike. This placed a burdensome responsibility on these people, many of whom would not have to bear the consequences of a prolonged strike, and who were hindered by their lack of a coherent, explicit political strategy.

There was a mass rally in Beckley. It was huge, overflowing. It was clear at the rally that miners wanted guidance in what to do. One of the legislators we had worked with got up and told them to go back to work, that they had won a great victory, that they should go back to work. And then he turned to me, and I'll never forget it, he said, "It's not worth the paper it's written on."

Nobody took a strong position at that rally. I didn't. I mean, I made a feeble attempt. I said, "This is not anywhere near what we needed." But it was just too heavy a responsibility on anybody to keep the strike going. Nobody was together enough to think of an alternative strategy. The thing was—"Stay out till the governor signs it." Of course he was going to sign. But they thought they'd stay out till then, make a big show of force.[67]

On Tuesday, March 11, Governor Arch Moore signed the black lung bill into law. About fifteen thousand miners in northern West Virginia had already returned to work on the midnight shift Sunday. Hours after Moore signed the bill, miners in the southern region began to show up for the evening shift. By daylight on Wednesday, West Virginia's coal industry was running once again. The black lung strike was over.

## Murder in the Union, Murder in the Mines

The ramifications of the black lung strike did not end with the miners' return to work. Although the movement had not achieved the broad compensation that activists had sought, it had put black lung on the agenda of the U.S. Congress. Debates over federal regulation of coal mining now explicitly addressed occupational health as well as safety, and respirable dust standards began to appear in proposed legislation. By the end of 1969, Congress passed the comprehensive Coal Mine Health and Safety Act, in which black lung compensation and prevention were prominent features. The federal compensation program established by this legislation would become the focus for a continuing movement centered on black lung.

The historical origins of the U.S. Coal Mine Health and Safety Act of 1969 are numerous. Most accounts point to the Farmington mine explosion and invoke the conventional wisdom that "dead miners have always been the most powerful influence in securing passage of mining legislation."[68] At first glance, dates and figures seem to confirm this observation, as nearly every federal health and safety law was preceded by disaster. Mine explosions have frequently precipitated press reports, bill drafting, and legislative discussion; however, their occurrence does not explain the content of reforms made in their wake. Indeed, the cries of outrage and accusations of negligence that invariably follow disasters have tended to obscure the underlying political and economic factors that shape health and safety regulation. The observation that it takes a disaster to produce reform simply suggests that concern for the life and health of workers is not built into the economic or political system. In order to explain the content of federal mine health and safety legislation, it is necessary to look beyond the sequence of disaster and reform, into the changing political economy of the coal industry.

The U.S. Coal Mine Health and Safety Act of 1969 was enacted after a period of rapid transformation in the production process and economic structure of coal mining. By the mid-1960s, a more streamlined, consolidated, and productive coal business was emerging from the rubble of an industrywide shakeout. Commercial coal operators' successful defense of lucrative markets, especially in electric utilities, persuaded their giant rivals in oil to switch their corporate strategy from competition to acquisition. In 1963, Gulf Oil entered the coal business by buying out Pittsburgh and Midway.[69] Two years later, George Love "stunned" Wall Street by selling off Consolidation Coal Company's assets to Continental Oil.[70] In 1968, Occidental Petroleum acquired Island Creek Coal Company, and Standard Oil of Ohio bought out the Enos and Old Ben mines. By 1970, three of the top ten coal companies, representing about one-sixth of annual production, were subsidiaries of oil.[71]

The entry of oil and other corporate money into the coal business brought new management, which lacked experience in the peculiarities of coal mining. For executives accustomed to overseeing international operations and a global work force, the diversity of production standards, the bewildering array of conflicting state safety regulations, and, above all, the informal, personalized deals with the UMWA president, appeared to be out of the nineteenth century. As one such executive commented:

> [At first] we tended to say to the coal people, "Just keep those dividends rolling in." But when we began to look closely, we found the most damnable things. . . . We found the coal company people calling the miners "the union employees" and everybody else "company employees." And then, after an actuarial study, we found that we could provide comparable pension benefits at a cost far lower than the 40-cents-a-ton royalty we're now paying. I've never seen anything like it. The coal people talk about Tony Boyle as if he was God.[72]

By the late 1960s, the structure of industrial relations in coal mining was overdue for change; personal collaboration had outworn its appropriateness to the industry and outlived its tolerability for the rank and file.

It was miners, however, who included workplace health and safety on the agenda of reform. The black lung movement, although confined to one issue and one state, came after nearly two decades of sporadic rank-and-file protest over hazardous working conditions. The strike was widely interpreted as evidence that discontent was on the rise—precisely at a time when its economic impact could be quite damaging. For coal operators, who were finally beginning to enjoy prosperity and stability, production stoppages and bad press over health and safety were more than temporary nuisances. As experienced miners retired en masse, analysts were predicting a severe labor shortage in the depopulated coalfields of Appalachia; many operators were seriously alarmed at developments that advertised the industry's disadvantages to young workers.[73] Moreover, it was unclear in the early months of 1969 what new upsurge might follow the black lung movement. The wildcat strike and militant demonstrations raised the unsettling possibility that the social revolt of blacks and students was spreading to the mainstream of the working class.

It soon became apparent that miners would not be appeased by a questionable compensation law in one state. Encouraged by the unrest in southern West Virginia, where the most populous districts in the UMWA were located, Jock Yablonski announced on May 29, 1969, that he would run as a reform candidate for union president against Tony Boyle.[74] A detailed battle plan for improving occupational health and safety headed his eleven-point platform. It called for expansion of the UMWA Safety Department, improved workers' compensation, greater health and safety research, and unspecified changes in collective bargaining contracts. This was not a radical or even a particularly far-reaching set of proposals; nevertheless, an insurgent candidacy under the general rubric of democracy and reform shook the union hierarchy at its base. As president of UMWA District 5 in western Pennsylvania, Yablonski, who had been involved since adolescence in coal mining and union politics, presented by far the most serious challenge to a UMWA president in over forty years. Indeed, his candidacy threatened not only the top leadership in Washington, but every single appointed official in the union's vast machine.[75] The Boyle-Yablonski contest coincided with the legislative debate over federal regulation of coal mine health and safety. It served one more notice on Congress that the miners' desire for improved occupational safety and health was not to be ignored.

Coal operators were strikingly divided in their response to the rank-and-file turbulence and the possibility of expanded federal regulation. The split within the industry in certain respects paralleled the division in the union between the supporters of the old regime and the forces pushing for reform.[76] Many operators who had been in the business for years, especially those in companies devoted exclusively to the mining of coal, longed for the days of solid handshake deals with John L. Lewis and viewed the autocratic union as an essential disciplinarian over the rank and file. They feared that the black lung movement and the Yablonski campaign presaged an era of anarchy, and they opposed most proposals for reform.

Other operators, predominantly though not exclusively in larger coal companies and corporations with diversified holdings, were intent on a new style of industrial relations. The old pattern had served the industry well, but now it granted the union president far too much discretion and made the contract dependent on the credibility and prestige of a single man for its legitimacy. More rational forms of authority were required. Change in the workplace as well as the union was necessary to enable the industry to attract sufficient new miners in a period of economic expansion and tight labor markets. As a writer in *Fortune* observed: "It is a matter of paramount concern to the operators that over the next ten years about half the labor force will have to be replaced, and these young miners are not going to tolerate hazardous working conditions. Nor will they buckle under to a union that acts dictatorially to impose the will of its leader."[77] A key group of coal operators thus supported federal coal mine health and safety reform but resolved to shape any legislation to their own advantage.

Behind-the-scenes negotiations between legislators and these reform-oriented coal operators were essential to the evolution of the Coal Mine Health and Safety Act. Congressional leadership originated not just among liberals in coal-producing states; other legislators aggressively pushed reform in the hope of establishing precedents for federal regulation of other industrial hazards and a national system of workers' compensation. Phil Burton of California, for example, authored the black lung compensation provisions of the act, and his efforts were critical to their passage. On the industry side of the negotiations, Consolidation Coal Company, now headed by John Corcoran and owned by Continental Oil, was again in the vanguard. One congressional staffer described the politicking:

> They picked out the most enlightened member of the industry and worked with him, and said, "Now John will you support the bill if we do X, Y and Z?" And Corcoran agreed to this. This is the real genius of a large part of this. Burton is a master legislator. And he knew that it was essential to have industry support here. He knew that large parts of the industry would not agree to the first cent being spent for safety or health, but that Corcoran was humanitarian, and was interested in this, and that John Corcoran would play some role in whether or not the bill was accepted by the President, and in whether or not other parts of the industry went along. If you could break the phalanx of opposition, it was essential to do that.[78]

Although Corcoran's conscious motive may have been humanitarian, more hard-nosed considerations also influenced his position. The previously mentioned labor shortage was one issue; it was also clear that the economic burden of any legislation would fall hardest on small, capital-poor companies. For big operators with mines in several states, more comprehensive, uniform regulation would help to standardize production practices and eliminate some of the small fry in the industry. Hence Corcoran and certain representatives from the BCOA consistently stressed that the act must apply uniformly to all underground mines and must be enforced uniformly, and that no fines for violations should be reduced because of a company's inability to pay (as was proposed). Before a Senate subcommittee, Corcoran argued:

> I think first we do need a new law, we need a strong law. But equally important, we need to see that that law is uniformly and vigorously enforced.
>
> I emphasize again that the law should apply to all coal mines. I think this is part of the problem today, the fact that many mines are not subject to the rigorous inspection that this proposed law would entail. . . .
>
> This committee has a better ability to effect a broadened safety posture for the industry than the States would have because there is no question, some States' laws are stronger and more vigorously enforced than others, whereas, Congress could apply a uniform law that I would hope then would be uniformly enforced. . . .
>
> My one concern here is to have a strong workable law that really will promote safety. As long as we have that kind of law and it is uniformly enforced and vigorously enforced, then this certainly will satisfy what I think is the requirement of the industry today.[79]

The legislation that finally emerged bore the unmistakable imprint of the large operators, but it also violated their wishes in a few important respects. The legislation was inclusive in coverage: "Each coal mine," "each operator" and "every miner . . . shall be subject to the provisions of this Act."[80] It removed a vestige of regulatory distinction between small and large mines by requiring that all electric face equipment be "permissible"—that is, designed not to produce sparks or otherwise contribute to the explosion hazard. However, in a key capitulation to the small operators, apparently required to win Republican votes for the entire package, legislators extended the period for installation of permissible equipment.[81] The act also spelled out precise, mandatory procedures regarding roof support, ventilation, combustible materials, communications, and other aspects of working conditions. It expanded the enforcement powers of the U.S. Bureau of Mines and gave federal inspectors the right to close mines where imminent danger existed. Although far more comprehensive than its predecessors, the 1969 law still emphasized reduction of explosion-related hazards, and its primary impact on safety seems to have been a reduction in fatalities from this source.[82]

Title IV of the act—"Black Lung Benefits"—established a temporary, federally financed program of compensation benefits for miners who were totally disabled by a broadly defined pneumoconiosis.[83] Widows of miners whose death was due to the disease were also eligible. A statutory presumption aided miners who showed evidence of pneumoconiosis and who had worked for ten or more years underground; their disease was presumed to be the result of their occupation, unless proven otherwise. Benefits were pegged to half the amount received by a disabled worker at grade GS-2 on the federal pay scale and were offset against payments from state-administered social insurance sources like workers' compensation and unemployment compensation. The program would be funded for three years out of the federal treasury and then revert to the state compensation system, provided that the state program met certain criteria established by the secretary of labor. In states with unapproved programs, the coal operators would be held liable for all new claims, and benefits would continue to be administered by the federal government. The entire program would end seven years after it was enacted.

It is safe to say that no one had any notion of how expensive the black lung benefits program would become. Originally designed as a temporary program to benefit retired miners and widows who would be ineligible for state workers' compensation because of statutes of limitations, the program became permanent due to the continued efforts of the black lung movement. Because no accurate prevalence data existed and no one had the foresight to anticipate activists' future successes in liberalizing eligibility criteria, predicting how many people would qualify for benefits was impossible. Democrats pooh-poohed Republican estimates of an annual bill of $384 million as a scare tactic designed to defeat Title IV. Little did they know that black lung benefits would drain the federal treasury of $1 billion per year.[84]

Although the compensation provisions aroused conflict within the U.S. Congress, it was the mandatory preventive measures that caused a flap in the industry. Unanimous opposition from the operators notwithstanding, the 1969 act placed the deep mines of the United States under the most stringent respirable dust standard in the world.[85] Title II required the operators of all underground mines to reduce respirable dust levels to 3.0 mg / $m^3$ air within six months from enactment. Three years after that, the dust standard would fall to 2.0 mg. The operators were also required to monitor dust levels through a periodic sampling program. Miners were to be given regular chest X-rays under a program administered by the secretary of health, education, and welfare; those with evidence of coal workers' pneumoconiosis had the right to transfer to a less dusty work area with no loss of pay. Some members of Congress insisted on these strict preventive measures as a quid pro quo for temporary federal financing of black lung compensation. In a larger sense, however, the compensation and prevention programs were all an outgrowth of coal miners' militant expression of discontent over occupational health and safety, and in particular, the black lung movement.

The U.S. Coal Mine Health and Safety Act of 1969 was passed by the Senate on December 18, 1969, nine days after incumbent Tony Boyle defeated Jock Yablonski for the presidency of the United Mine Workers. Balking at what he considered the "budget busting" black lung compensation provisions, President Nixon initially declared his intention to veto the legislation. On December 30, however, he signed the Coal Mine Health and Safety Act into law, hours after receiving reports that twelve hundred West Virginia coal miners had walked out and were calling for a nationwide shutdown of the mines.[86] One day later, on New Year's Eve, Jock Yablonski, his wife, Margaret, and daughter Charlotte were murdered in their Clarksville, Pennsylvania, home. The trail of blood wound back through the Appalachian ghetto in Cleveland, through the union's machine in eastern Kentucky and Tennessee, to the UMWA headquarters in Washington, D.C.[87] At the Yablonskis' funeral, miners mourned the death of their leader, then founded a new organization and pledged that the movement itself would not die.

## Conclusion: The Limits of Reform

Repression and concession are the primary responses of the state to protest from below, according to most analysts of social movements.[88] Within the framework of these two alternatives,* the state's response to the black lung movement typically has been recorded as one of concession extracted by the unified economic power of striking coal miners. The movement won a great victory by forcing the West Virginia legislature to extend workers' compensation coverage to victims of black lung disease. As one writer described it: "The strike . . . ended only after

*A third alternative is simply to ignore the protest.

the legislature acceded to miners' demands and enacted one of the most enlightened worker's compensation laws in the country."[89]

In reality, the interaction between movement activists and the political machinery of the state was immeasurably more complex. The point is not simply that the black lung strike ended with a more compromised, symbolic victory than most observers have assumed. In a broader sense, from the first moment that miners formulated their goal in terms of existing political alternatives—workers' compensation coverage of black lung disease—they began a relationship with the established processes of reform that gradually and subtly redefined the target of their anger, the goal of their activism, and the political meaning of their discontent.

The righteous anger that infused the black lung movement was rooted in the miners' experience as workers in the postwar years, a time of deteriorating working conditions, rising dust levels, insecure jobs, and rank-and-file powerlessness. Discontent focused above all else on the coal operators; they were the ones who "wore out everything [the miners had]." Black lung disease was a source of physical pain and disability, but it also assumed more abstract meaning as a symbol of social suffering and loss. Compensation was payment for a medically verifiable occupational lung disease, but it also represented retribution for a lifetime of exploitation. Neither the West Virginia legislature nor the compensation bureaucracy was the source or the initial target of miners' and their families' anger. The black lung movement arose from the class-based discontent of workers who sought to settle accounts with the coal operators for the suffering they had caused.

However, once miners and other activists settled on the single demand for black lung compensation and the strategy of legislative reform in West Virginia, in significant ways they lost control over the meaning of their goal. Translating the demand into a form compatible with established political procedures (i.e., a bill for passage by the West Virginia legislature) required expertise in areas where miners had no training: law and medicine. They were forced to rely on sympathetic professionals who could draft legislation and then advise them as to which compromises were acceptable. As the West Virginia legislature debated and revised the black lung bills, miners in the galleries looked on at a process that they could influence but not control. Political compromises as well as the formal rules and schedule of the legislature shaped the content of what the miners accomplished: once the session adjourned for the year, miners had few alternatives to acceptance of the legislation, especially when some of their advisors asserted that they had "won a great victory."

At a deeper and less evident level, the workers' compensation system through which activists sought a just retribution inherently thwarted their goal. Workers' compensation evolved during the Progressive Era as a political response to mounting damage claims from workers and their families who attempted to hold corporations responsible for death and injury in the workplace.[90] The state-based programs theoretically substituted speedy and equitable—but limited—compensation for the uncertain and expensive process of suing for negligence. Under workers' compensation, most employers must pay premiums into a state fund or insure themselves at a specified rate against the disability claims of their employees. The system places a cap on corporate liability by eliminating workers' right to sue their employers for damages. It quantifies physical loss in a standardized fashion: "The loss of a great toe shall be considered a ten percent disability."[91] Workers' compensation thus makes occupational death and injury an insurable expense, a predictable cost of doing business.

Even within these structural guarantees of limited corporate liability, most workers who are completely disabled find it difficult to win a permanent compensation award.[92] Those with occupational disease face the frequently insurmountable legal requirement that they prove its origins in the workplace. According to a 1980 report by the U.S. Department of Labor, 3 to 5 percent of workers "who report severe disability from illnesses perceived to be occupationally-related received workers' compensation."[93] Because the system implicitly equates disability with the incapacity to perform wage labor, work-related problems like sexual impotency are not necessarily covered by workers' compensation.[94] Even those who are severely and clearly injured from an accident on the job can rarely prove "permanent and total disability," the legal requirement for a lifetime award. Typically, they receive limited medical coverage and temporary compensation benefits, after which they face a lifetime of poverty. No state grants even "permanently and totally disabled" claimants the equivalent of their lost earnings for life; benefits are calculated as a fraction of their former wages and/or the average wage statewide. According to one estimate, compensation replaces on average less than 25 percent of a worker's wages.[95]

The black lung strike of 1969 publicized the problem of unrecognized occupational disease and alerted other workers that their jobs might be harmful to their health. It inaugurated an era of rank-and-file agitation over workplace health and safety that continues to this day. The retribution that activists hoped to exact from the coal operators never materialized, however. Insurance premiums for miners' disability due to pneumoconiosis became another calculation on the corporate balance sheet. Legislative compromises over wording of the black lung compensation legislation narrowed eligibility for benefits and further restricted the substance of the movement's apparent victory. One year after the new black lung law took effect in West Virginia, a grand total of eleven coal miners were receiving lifetime awards for permanent and total disability due to occupational pneumoconiosis.[96]

The black lung strike also led to passage of Title IV of the U.S.

Coal Mine Health and Safety Act of 1969. As measured by the volume of claims awarded to miners and widows, this program was an astounding victory that far surpassed the legislative reforms in West Virginia. Yet, it too averted the moral and financial reckoning with coal operators that black lung activists had sought. Indeed, the federal compensation program represented an indirect subsidy to the coal industry; it paid for the work-related disabilities of miners with general tax revenues from the U.S. Treasury. Federal black lung benefits allowed the industry to escape responsibility for the human consequences of mechanization and to place the financial cost of compensation on the taxpayers.

The ambiguous victory achieved by black lung activists was not the product simply of political deals and compromises. More subtle processes were also at work. In the formulation and implementation of the new legislation, the miners' initial demand for retributory black lung compensation was refracted through the lens of a scientific and technical knowledge that filtered out its intensely political essence. For miners and their families, black lung was the terrible product of historical exploitation, the tangible evidence of their collective entitlement to redress. In the view of scientific medicine, however, "black lung" was an unscientific, lay term for a collection of specific disease entities, all of them associated with distinct physical hazards, most of which were nonoccupational in origin. Entitlement to compensation had little to do with miners' physical self-evaluation, much less their social history as workers. Eligibility could be established only through individual, case-by-case diagnosis of a single clinical entity, coal workers' pneumoconiosis, and the quantitative assessment of any associated disability.

A similar scientific rationale underlay the black lung prevention provisions of the U.S. Coal Mine Health and Safety Act. For activists, the disease ultimately arose out of antagonistic social relations between coal miners and operators, which had permitted a gross deterioration in working conditions during mechanization of the mines. From a technical perspective, however, CWP was "caused" by the inhalation of minute particles of respirable dust. Such a physical hazard was amenable to engineering controls: respirable dust levels could be permanently reduced by increasing ventilation, placing water sprays on equipment, and monitoring the results through regular air sampling. These activities are, of course, essential preventive measures; the point is that, unless they recognize the human beings whose power, interests, and actions determine actual conditions in the workplace, they are empty academic prescriptions.

In other words, the scientific and technical knowledge that underlay these reforms was itself laden with assumptions that redefined the political meaning of what miners sought. This knowledge base transformed an explosive social problem into a physical phenomenon, a collective experience into an individual diagnosis, a class-based injustice identified by miners and their families into a medical / technical problem to be assessed by those with professional expertise. Scientific and technical knowledge mediated between the demands of discontented workers and the policy response of the state. Apparent neutrality and objectivity lent such knowledge formidable ideological power, even as its implicit structural assumptions complemented and upheld the prevailing social order. Individualism, physical / biological reductionism, and a scientism that required the instruments and expertise of professionals to legitimate the very existence of a social problem—all functioned to obscure the relations of power that conditioned the production of disease.

Enactment of the West Virginia compensation law and U.S. Coal Mine Health and Safety Act reshaped the field of action in which miners and their families sought to grapple with the black lung issue. Those who participated in the movement that continued after 1969 formulated their goals and activities in reference to this legislation. The arena of conflict shifted out of the workplace, away from the confrontation between miners and operators, and onto the inscrutable processes of remote government bureaucracies. The federal compensation program took the heat off the industry and put it on agencies and administrators in Washington, D.C. The preventive measures mandated by the 1969 act generated a complex of regulations and a hierarchy of bureaucratic personnel that became an important cause of—that is, scapegoat for—the industry's occupational hazards. Coal miners lost their lives and health in this new era because the Bureau of Mines was slow moving and complacent in its enforcement of federal regulation.

Ironically, both reforms sanctioned the respective authority of physicians and the coal operators—those whom miners had fought bitterly to achieve recognition of black lung in the first place. The compensation law in West Virginia and the black lung benefits provisions of the 1969 act assigned to "licensed physicians . . . of good professional standing"[97] the power to diagnose occupational pneumoconiosis and ascertain any associated respiratory disability in each claimant. Although nonmedical findings, such as the number of years of employment in the mines, were also essential to establishing a "valid" claim, eligibility often hinged on the medical reports of physicians.

Similarly, the coal operators were assigned responsibility for preventing black lung. Control of dust sampling, the essential process for monitoring the extent of this respiratory hazard and, by implication, the effectiveness of any control measures, was placed in the hands of the operators. Throughout the text of the 1969 act, coal miners were explicitly *not* granted power or responsibility to ensure their own health and safety. The underlying issue of workplace control emerged most clearly in congressional debate over a proposal to fine miners as well as operators if they violated a health or safety regulation. The union and the industry opposed this provision; both argued that control of the workplace, the work process, and the work force is the right of management, and therefore health and safety is management's responsibility. UMWA president Boyle testified: "These operators . . . own these mines and the United Mine Workers membership will not abridge the right or the authority of the coal operator in running those properties."[98] In sum, these reforms enlarged the scope of government regulation of coal mining, but they did so in ways that upheld existing relations of power, control, and authority.

The effectiveness of these reforms is a matter of dispute. Since passage of the 1969 act, the fatality rate in underground mines has in fact dropped, due in large part to a reduction in deaths from mine explosions. However, with the exception of the 1974–76 period, the rate of disabling injuries in underground coal mines has actually worsened. Although changes in methods of reporting and calculating make comparisons of mine safety statistics difficult, it appears that in 1979 and 1980 the rates of disabling injuries underground were among the highest since World War II.[99]

As for dust control, in 1974 the secretary of the interior confidently reported that 94 percent of all underground mine sections were in compliance with the strict respirable dust standard of 2.0 mg / $m^3$ air.[100] Other evidence suggested a very different story. A study of dust conditions conducted in the same year by the U.S. Bureau of Mines concluded that miners were being exposed to "grossly excessive amounts of respirable dust."[101] One year later, a study by the General Accounting Office found so many weaknesses in the dust sampling program that the researchers were forced to conclude: "Current procedures [make it] virtually impossible to determine how many mine sections are in compliance with statutorily established dust standards."[102]

Among the weaknesses they cited were a minimum margin of error in dust measurements of ±32 percent, "when taken by trained scientists using meticulous care," and operator control of the sampling program.[103] Allowing the fox to guard the chicken coop has led to predictable abuses. In 1979 and 1980, three of the largest U.S. coal producers, Consol, Westmoreland, and Pittston, were hauled into court on charges of deliberately falsifying dust samples. Westmoreland and a Pittston subsidiary were convicted; the latter was fined $100,000.[104]

Even the state and federal compensation programs brought equivocal benefits. Black lung claims were judged according to the restrictive medical viewpoint that limits miners' occupational respiratory disease to coal workers' pneumoconiosis; as a result, many claims were denied. Physicians now stressed that cigarette smoking was far more pernicious than occupational hazards in its effects on miners' lungs. In 1970, after a year of examining claimants under the new compensation law, the three physicians who composed the Occupational Pneumoconiosis Board of the West Virginia workers' compensation system recommended the following preventive action: "A vigorous and persistent campaign should be carried on by all parties interested in preventing lung disease in coal miners to point out the hazard of cigarette smoking and to encourage all coal miners to become nonsmokers."[105]

Denial of miners' and widows' claims for federal compensation raised a storm of protest in the coalfields. Black lung nominally had been recognized as coal workers' pneumoconiosis, but this scientific construction did not correspond with the miners' own experience of disease. In the years that followed their first apparent victory, activists sought to reaffirm the goals of their movement. In hearings before administrative law judges, miners, widows, and lay representatives argued the legitimacy of individual claims for black lung benefits. In confrontations with federal program administrators, they contested the criteria for eligibility and attempted to influence the bureaucracy that their actions had created. In public hearings and legislative debates, they faced off with physicians over disability standards, X-ray evidence, and other increasingly esoteric questions of diagnosis and etiology. At the heart of this renewed movement was an intensely political struggle over who would determine federal policy on compensation, who would set the criteria for work-related disability, and who would control the definition of black lung.

# 5

## Carry It On

Anyone who has walked on a warm spring day through the streets of Welch, West Virginia, or any of the hundreds of coal towns in central Appalachia, would probably conclude that black lung is a pervasive disease, easy to recognize, and severely disabling. The older miners who gather around the park benches and storefronts walk slowly, stop frequently to catch their breath, and periodically hunch over in convulsive fits of coughing. Many are barely fifty years old. Yet, these miners represent only a portion of the disease-stricken, those who are least disabled. In countless homes and hospitals lie thousands of other miners so disabled by black lung that they cannot even get out of bed.

The prevalence of black lung disease is unknown. Conflicting data and viewpoints abound. In 1969, the U.S. surgeon general estimated that 100,000 coal miners were afflicted with black lung.[1] Four years later, over 150,000 miners had received federal compensation benefits for total disability due to the disease.[2] Epidemiological studies of coal workers' pneumoconiosis have found overall prevalence rates in the work force that vary from 10 to nearly 30 percent;[3] moreover, the prevalence of disease may fluctuate over time. Demographic changes in the work force (e.g., toward a younger population with fewer years of dust exposure), changing work practices and levels of dust, variations in the eligibility standards for pensions and black lung benefits (which influence older miners' decisions about leaving the work force)—all affect the prevalence rate. To complicate matters further, most of the recent, large-scale epidemiological studies of coal miners' occupational respiratory disease have focused on coal workers' pneumoconiosis, which, some physicians, assert is not the sole work-related lung disease among coal miners. If other potentially work-related conditions (e.g., chronic bronchitis and emphysema) are included, the overall prevalence of "black lung" is much higher, but unknown.

When the Social Security Administration began implementing the federal black lung benefits program in 1970, the agency judged claimants' eligibility according to narrow criteria that reflected the increasingly dominant medical understanding of black lung. Despite the broad legal definition of disease expressed in the 1969 act ("a chronic dust disease of the lung arising out of employment in an underground coal mine"),[4] miners without X-ray evidence of pneumoconiosis were automatically refused benefits. Two years into the program, over 347,000 miners and widows had applied for compensation. In West Virginia, 55 percent of all claims were denied, and in Kentucky, the denial rate was far higher: social security turned down 68 percent of all claimants. Nearly two-thirds of the denials nationwide were based on claimants' lack of X-ray evidence of coal workers' pneumoconiosis.[5]

The process of attempting to establish eligibility for compensation usually began with a visit to the local social security office, where claimants filled out their application for benefits.[6] Miners without a recent chest X-ray were sent to mobile units, where technicians took a film of their lungs. Many claimants did not make it beyond this first stage; based on a single chest X-ray and no other medical evidence, they were turned down for compensation. The few whose X-rays revealed complicated coal workers' pneumoconiosis were automatically granted benefits based on an irrebuttable, statutory presumption of total disability. Those whose X-rays indicated simple pneumoconiosis faced a second legal hurdle: they had to prove, typically with evidence from a pulmonary function test, that they were totally disabled from pneumoconiosis. The standard for "total disability" was rigidly quantified: claimants had to meet a precise numerical cutoff established by the Social Security Administration. Those who did so qualified for compensation.

Frequently, claimants met one of the eligibility criteria but failed to meet another, and this failure was used to disqualify them. Miners with evidence of respiratory impairment as measured by a pulmonary function test or blood gas test were turned down because their X-ray did not reveal pneumoconiosis. Others, whose X-rays clearly and consistently indicated CWP, were denied because their pulmonary function test did not meet the numerical standard for "total disability." The Social Security Administration then began sending off claimants' X-rays for rereading, and miners who had believed they were eligible for compensation were denied after physicians reinterpreted their chest films. Widows of miners who died before the medical profession even recognized the existence of CWP faced special problems, as their husbands' medical records rarely mentioned an occupational lung disease. Even those with physicians' reports of serious respiratory impairment were turned down if their husband's death occurred, for example, in a car accident, and therefore was not due to pneumoconiosis.

For those who lived each day with respiratory disease—miners and their families—these eligibility standards for compensation appeared completely arbitrary and incomprehensible. Miners who had spent thirty or forty years underground, whose breathing was a series of audible rasps, whose hacking coughs regularly produced inky black sputum, whose retirement did not include the hunting and fishing they had dreamed of, but, rather, short walks between the bedroom and the kitchen—all over the coalfields, such miners were denied compensation. The eligibility criteria for benefits seemed unrelated to these miners' palpable experience of disease and disability. The news that they were not "totally disabled," or that their X-ray did not reveal pneumoconiosis (which they equated with black lung), was not just absurd, it was personally insulting. This disjunction went beyond practical considerations—that is, money in the form of compensation benefits at a time of financial need—it encompassed a moral and political tension. To deny such miners compensation was to deny the legitimacy of their disability, the reality of their pain.

Claimants initially directed much of their anger and resentment over the compensation program at an accessible human target: the staff of the local social security offices. Prepared for neither the volume nor the complexity of the black lung claims, the local offices became swamped with letters, applications, and medical records, their waiting rooms filled with frustrated claimants. During the program's first month of operation alone, an overwhelming one hundred thousand claims flooded the agency.[7] The staff of the local offices was responsible for assembling the evidence necessary to evaluate each claim. Social security workers tended to retrieve easily available medical reports and work records from other institutions—for example, workers' compensation or the UMWA—but rarely tracked down more obscure records or informed claimants of their right to obtain additional medical evidence at the agency's expense. Hence, claims from states like Pennsylvania, where CWP had been recognized as compensable five years earlier, had a relatively high approval rate in part because a system of medical records on miners' occupational lung disease already existed; by the same token, claims from Kentucky and West Virginia had a higher rate of denial.[8]

When, in the fall of 1970, claimants first received form letters indicating they had been denied compensation, many returned in bitterness to their local social security offices. In many instances, they obtained little assistance from personnel, who did not understand the medical technicalities involved in establishing eligibility themselves. Some miners and widows asked to see the contents of their claim file in order to reach an understanding of why they were turned down; in several offices, they were refused. Others were made to feel that black lung compensation was an unwarranted form of welfare, a handout for which they should be ashamed to ask.[9] Early in 1971, miners and widows in southern West Virginia expressed their frustration with the agency in a leaflet that they distributed at the social security office in Logan:

> After waiting a year [for news on our black lung claims] we get a form letter from the Government saying we are not going to get paid. The letters are all the same and don't say a thing. When we go to Social Security for help we are insulted, ignored and get the run around. The Social Security people won't tell us why we are turned down. They won't let us see our files. They won't help us get new evidence. All they do is try to discourage us and get us out of their office. This had better stop.[10]

The federal compensation program became the target of a continuing controversy over black lung. During its second phase, which began in 1970, many of the movement's features changed. The focus of participants' anger shifted to the national level, away from southern West Virginia and the coal operators, onto an existing government bureaucracy. The Black Lung Association developed into a thriving collection of county-based organizations that provided claims assistance as well as political leadership in pressing for compensation reform. The movement spread into southwestern Virginia, eastern Kentucky, and other scattered locations in the coalfields; although participation broadened geographically, it was concentrated more and more among retired and disabled miners, their wives, and miners' widows—not among those still working in the mines. Power resources and pressure tactics deployed by the movement were commensurate with this changing social base; there were no more massive wildcat strikes over black lung, but, instead, there were marches, demonstrations, and media events designed to embarrass agency officials.

Despite these dissimilarities, at the heart of the movement was a continuing struggle with physicians and policymakers over the definition of black lung and the eligibility standards for disability compensation. "Black lung" acquired complex and contradictory meanings, as activists learned to challenge with their own experience the finer points in the dominant medical construction of coal workers' pneumoconiosis. Their political successes in reforming the compensation program yielded an ever-changing legal definition of the disease. Meanwhile, physicians themselves became more visibly and stridently divided over what constituted occupational respiratory disease among coal miners.

## The Definition of Disease

The position of the Social Security Administration initially coincided with the dominant medical view of black lung: the agency compensated only miners with X-ray evidence of pneumoconiosis. The scientific rationale for this approach is that pneumoconiosis is a disabling respiratory disease exclusively related to dust inhalation. It is not associated with cigarette smoking or other nonoccupational sources. Pneumoconiosis may be diagnosed in the living miner only through chest X-rays; indeed, the disease presumably progresses and impairs in cumulative stages that are delineated according to the radiographic image of the lungs. "Simple" pneumoconiosis, or the early stage of disease, is deemed compatible with health; respiratory impairment in miners with simple pneumoconiosis is presumed to be related to nonoccupational sources, especially cigarette smoking. "Complicated," or advanced, pneumoconiosis is associated with severe disability; the estimated 2 or 3 percent of miners with this stage of the disease are considered genuinely disabled from their work and therefore deserving of compensation.[11]

Critics of this approach within the medical profession charge that the exclusive emphasis on X-ray evidence of CWP arbitrarily restricts the problem of occupational lung disease among coal miners. In the first place, the system for classifying X-ray findings regarding pneumoconiosis was never intended as a correlate for disability.[12] Some miners with simple CWP have extreme functional impairment of the lungs; thus, any rigid and facile demarcation between simple pneumoconiosis ("health") and complicated pneumoconiosis ("total disability") is unjustified.[13] Moreover, X-rays are imperfect diagnostic tools, especially for the early stages of disease.[14] Most importantly, dissenting physicians insist that coal miners can contract many forms of lung disease in the workplace. Bronchitis, emphysema, impaired gas exchange—all are associated with dust inhalation, all can be severely disabling, and none (except advanced emphysema) is evident on a chest X-ray. Epidemiological studies have found a higher prevalence of these problems among coal miners than among nonminers, and in some cases they have correlated dust exposure with the incidence of disease.[15] These conditions are not exclusively occupational in origin, but that is no reason to eliminate them automatically from consideration as compensable.

This critical standpoint was not limited to activist doctors in the coalfields. In 1971, a team of twelve physicians from all over the United States, several of whom were pulmonary specialists, gathered in Beckley, West Virginia, to examine thirty coal miners who had been denied black lung compensation. Their findings and recommendations indicted the existing eligibility standards and administrative policies of the Social Security Administration, which, the physicians charged, were "unduly and unnecessarily restrictive."[16] The physicians issued a joint statement in which they concluded:

> a. [That] there is a diversity of pulmonary diseases and conditions associated with coal mining for which the rigid definition of "pneumoconiosis" (possessing as its *sine qua non* a radiologic lesion) is not tenable;
> b. That disability from work-associated respiratory disease after appropriate review be compensated;
> c. That criteria for eligibility for all work-associated pulmonary disease be based upon *functional impairment* rather than solely upon anatomic or dadiologic criteria.[17]

The Social Security Administration adopted and defended a restrictive definition of compensable disease in part because of its own organizational interests. The black lung compensation program represented a major, unexpected drain on the agency's resources, particularly its staff time and budget. Organizational considerations dovetailed with the conservative medical viewpoint on CWP, and with the political stance of the Nixon administration, which had opposed the federal compensation program from the start. Thus, the agency utilized the X-ray requirement as a convenient and relatively inexpensive screen to eliminate many claimants from consideration at an early stage in the process. Unlike, for example, blood gas studies, which are costly and require extensive laboratory equipment not widely distributed through the rural coalfields, X-rays are cheap, quick, and accessible. Rigid adherence to a definition of disease that exclusively required X-ray evidence for diagnosis reduced the complexity and expense of administering an unwieldy (and politically unwanted) program.[18]

The exclusive focus on X-ray-diagnosed CWP also had deeper meaning, in that it reflected fundamental tendencies in clinical medicine that many, including physicians, have criticized and lamented. Since at least the turn of the century, there has been a marked trend in Western medical practice toward the substitution of apparently objective tests and measurements, made possible by a growing array of technological devices, for the careful clinical examination of the patient.[19] Neither the patient's report of symptoms nor the physician's perception of signs retains its former legitimacy; both must be constantly validated through medical testing. Enthusiastic supporters of "laboratory medicine" point to the precision and objectivity that it lends to the diagnostic process. Critics call this claim into question by citing studies that indicate wide discrepancies in the results and interpretations of X-rays, certain laboratory tests, and other "objective" procedures. For example, research has documented significant disagreement in the analyses of the same X-ray by several radiologists. The type of film and the technique used in taking the X-ray can also influence its interpretation.[20]

The example of X-ray-diagnosed CWP suggests an additional problem, one more profound than dependence on diagnostic tools of questionable reliability or validity: technological devices tend to become all-powerful arbiters of the existence of disease; indeed, they shape the very definition of disease. Coal workers' pneumoconiosis exists independent of the patient's symptoms, in the disembodied opacities of a chest film; diagnosis does not necessarily even require a personal encounter between physician and coal miner, much less an extensive clinical examination. In the context of the black lung compensation program, this technological construction of disease led to a ludicrous and tragic situation throughout the coalfields: disabled miners battled for breath while physicians four hundred miles away viewed X-rays of their lungs and pronounced them free of occupational respiratory disease.

In contrast to the various medical concepts of black lung, miners and their families posited a radically different understanding of disease. Although when cast in scientific medical terms their perceptions most closely resembled the broader definition of dissenting physicians, their understanding of black lung exceeded the boundaries of medical science in important respects. For movement participants, the medical definition of black lung as a disease (or diseases) effecting specific physical changes in a single organ of the body had little meaning; their own experience of black lung involved a transformation in their entire way of life. One fifty-six-year-old miner, disabled by black lung since the age of forty-eight, described how the disease had affected his life:

> Black lung is a cruel disease, a humiliating disease. It's when you can't do what you like to do, that's humiliating. I had to lay down my hammer and saw, and those were the things I got the most pleasure out of. The next thing I liked to do was work in my garden; now my garden's the biggest weed patch in Logan County. There were times in 1971 when I was still working that it was difficult for me to get to the bedroom when I was feeling bad. Now, of course, that's humiliating.[21]

Miners' analysis of the physical agents associated with black lung was based on their intimate, practical knowledge of the workplace. They were workers who day after day had been exposed to the numerous physical hazards of coal mines and who down through the years had witnessed the gradual deterioration in their own health: the increasingly severe colds, the bouts of bronchitis, and a diminished ability to breathe. They argued that many features of the workplace had damaged their health: working daily in water over their ankles, working in clouds of dust, working around the fumes from cable fires, working shifts. Disease was a generic process that began on the first day a miner started to work in such a health-destroying environment, not some randomly captured "spots" on the film of an X-ray.

> I worked in the cleaning plant, an outside job [above ground]. I had four conveyors to bring to the storage bin. I had, I'd say, 16 holes

> in this galvanized pipe, two rows, that's 32 holes in all, little tiny holes, to keep down the dust. I stood many a time across from that conveyor and somebody'd be on the other side, and all you could see was their cap lamp. And that's in the cleaning plant; that's outside! That's not even at the face. In the Black Lung Association, we're asking due compensation for a man who had to work in the environment he worked in.[22]

Activists' understanding of the causes of black lung derived from their social experience with the coal industry, as workers, as widows of men killed in the mines, as residents of coal towns where "there are no neutrals,"[23] not even scientists. Although respirable dust might be the agent of CWP, the ultimate cause of the entire disease experience was economic:

> Where do we get the black lung from? The coal companies! They've had plenty of time to lessen the dust so nobody would get it. It's not an elaborate thing to keep it down: spray water. They just don't put enough of it on there. They don't want to maintain enough in materials and water to do that.[24]

> Should we all die a terrible death to keep those companies going?[25]

This cultural, implicitly class-based understanding of black lung conflicted with the dominant medical view on virtually all counts. Miners and their families pushed the definition of disease beyond a purely physical event to encompass the social relations of the coalfields. They integrated economic exploitation and physical harm from the vantage point of people who experience these problems as coextensive, internally related, part of the same world of social / physical experience. Their implicit integration of social and physical phenomena challenged the divorce between human experience and the physical world that is the hallmark of "hard," "objective" science. Their broad, experiential definition of disease contrasted vividly with the prevailing medical construction of a single clinical entity, CWP, and the associated emphasis on a single diagnostic tool, the X-ray. Their collective insistence on the reality of their own disease experience defied the medical tendency to individualize health problems as separate, clinical cases and to denigrate patients' perceptions of their own condition. The scorn with which many physicians viewed miners' understanding of black lung was matched by activists' hostility toward the medical establishment—an attitude rooted in the history of company doctors and reinforced in the present by physicians who insisted that miners' respiratory impairment was due primarily to cigarette smoking.

During the many years that the black lung movement persisted, it acquired different purposes and meanings in the eyes of various participants. For many, the movement was a straightforward effort to obtain deserved compensation. For some, it was a means of organizing a rank-and-file insurgency capable of overthrowing the hierarchy of the United Mine Workers. For others, it was a vehicle to challenge the structure of class relations that produced disease in the first place. Running through these diverse goals and perceptions was the powerful symbol of black lung itself. Black lung's complex and unifying meaning—as physical disease, as social exploitation, as affirmation of entitlement to redress—was one of the great ideological strengths of this movement.

## The Black Lung Association

The murder of the Yablonskis in December 1969 temporarily chilled rank-and-file activism all over the coalfields. Those who had visibly participated in the black lung strike and the Yablonski campaign lived in fear that they, their spouses, and their children might be next. One activist recalled: "Everything kind of went underground. Everybody started carrying guns, getting floodlights for their yards, running around with big dogs, that kind of thing. It was a very scary time."[26]

Suddenly, in the summer of 1970, another wildcat strike broke out in southern West Virginia. Organized by disabled miners and widows and led by a militant black miner named Robert Payne, the strikers demanded that UMWA president Boyle meet with them concerning pensions and hospital cards. Pickets spread the walkout into eastern Kentucky, Virginia, and Pennsylvania, idling thousands of miners. Men in wheelchairs and widows with children appeared on the picket lines to support the strike; in one county, ten women banded together to shut down the mines. This wildcat persisted on and off for almost two months, and turned into a ground war between Boyle loyalists and the forces in support of reform. UMWA district officials stumped through coal camps attempting to persuade miners to go back to work; pickets brought them out again. As tensions escalated, many miners began carrying guns. Federal marshals swept through the coalfields seeking the individuals named in a temporary restraining order against the strike. One disabled miner quipped, "In the old days, the coal companies had guards. Now they have the courts."[27]

The strike eventually petered out as Boyle stonewalled on the protesters' demands and a district court judge threw Payne and other leaders in jail for contempt of court. Although it failed to win its stated goal, this strike was of great symbolic significance for coal miners and their families. It cut through the atmosphere of fear, renewed the spirit of defiance, and reactivated the organization of a movement.

In southern West Virginia, one spearhead for the renewed organizing efforts was a group of young political activists, all veterans of the war on poverty, who had founded a nonprofit organization, Designs for Rural Action (DRA). Following the congressional defeat of independent, federally financed community action programs in 1968, these organizers put together DRA as "a vehicle for continuing to do community action-type work. And so we could also support ourselves and not have to work for anybody else'':

> There were about five or six people who set it up; all of us were in the Appalachian Volunteers, except for one. We set it up with the idea that it would last five years, that that was long enough for any institution to last. That if you set up any institution for change, it's likely to get—harder, more rigid. And it lasted four and one-half years. . . . In terms of raising hell and causing trouble, it was the most effective thing I ever worked with.[28]

As veterans of the war on poverty, the founders of DRA focused initially on ''poor people's'' issues such as welfare rights rather than on the struggle within the United Mine Workers. However, the black lung strike in the winter of 1969 drew their attention to the economic power and latent discontent of coal miners. During the summer of 1969, DRA staff members began meeting with the remnants of the Black Lung Association, and by the following year, they had decided to concentrate their energy and resources on reactivating the black lung movement. Their strategy was to use the dissatisfaction with the federal compensation program as a means to revive the momentum of activism and keep together a network of dissidents; their ultimate goal was to build the organizational base that could overthrow Tony Boyle and reform the United Mine Workers of America. One of DRA's chief organizers recalled:

> The black lung strike, the strike in 1969, was a real spontaneous kind of thing, as far as I know. It never developed any organizational structure to keep continuing. There was no communication structure; the organization was practically gone. So we decided in DRA that our main priority was going to be the kind of thing we knew how to do—get people together, get them enough back together on black lung. But it was more using the black lung thing to get at the union.[29]

DRA's underlying motive of union reform originated in the

widely held perception that the corporate practices of the coal industry were the ultimate source of most social problems in central Appalachia. In this coal-dependent region, the sole force that could challenge the industry on a consistent and effective basis was the collective organization of coal miners—the United Mine Workers of America. One organizer in southern West Virginia explained: "I saw the whole thing as, all of the problems in the area—black lung was one of them—and the source of all of them was the coal industry. The union was the one thing that could stand between miners and the coal companies. You had to reform the UMW for it to put pressure on the operators to clean up the mines. Black lung was like strip mining; that was just the issue at the time."[30] DRA staff and rank-and-file leaders chose to organize under the rubric of black lung (rather than directly for union reform) because the issue was neither divisive nor controversial among miners and their families. Frustration and discontent with the federal compensation program guaranteed a large following of people who would not have to repudiate the Boyle hierarchy in order to join the Black Lung Association: "The black lung thing everybody could agree on. We didn't have to directly attack Boyle; you could show by contrast that the union wasn't doing anything."[31]

Countywide Black Lung Association chapters became the local organizational vehicles for this strategy. The chapters served a triple function: they offered assistance through lay advocates for individuals pursuing federal black lung compensation claims; they pressured the Social Security Administration and the Congress to reform the compensation program; and they organized a network of dissident UMWA members. In the summer and fall of 1970, several rank-and-file miners, VISTA workers, and independent young organizers began to organize Black Lung Association chapters throughout southern West Virginia. Many trained as lay advocates and attracted participants to their meetings and other activities through the service they provided. Some coal miners and union members were able to emerge as insurgent leaders in part because of the personal following they developed among individuals who needed assistance with their black lung claims.

The DRA office in Charleston served as a meeting place and focal point for the growing network of dissident BLA leaders. The small storefront was the hub for all sorts of progressive causes, from draft resistance to black lung, from abolition of strip mining to union reform. With its Goodwill furnishings, overflowing ashtrays, and continual stream of energetic young "organizers" in pursuit of one or another political goal, this office was like hundreds of others scattered across the country during the 1960s. Today, the building in which it was housed has been razed, replaced by a hospital.

The DRA staff, augmented occasionally by Antioch College students and various free-floating organizers, promoted communication among the leaders by issuing an erratic series of newsletters on local BLA chapter activities, along with the somewhat more regular *Black Lung Bulletin*. The monthly *Bulletin* covered the insurgent struggle within the United Mine Workers as well as developments related to black lung compensation.[32] The staff also orchestrated numerous media events designed to maintain the visibility of black lung as an issue and to project the appearance of a large and powerful organization. Commented one organizer:

> The Black Lung Association was kind of a mythical organization. A good example is something Gibbs [Kinderman, DRA staff person] was a master at doing. For some reason, Richardson, the secretary of HEW, was in Charleston for a meeting and Arch Moore was there and Gibbs got four coal miners and three signs and two TV stations and got a press report saying, "Miners protest over black lung." He knew how to size up a situation and get the best out of it. And he taught a lot of us; we did a lot of that. We could shut down the mines with a half-dozen guys. It was sort of guerrilla tactics. There was a

whole lot of people out there who sympathized with what you did, but they weren't in the organization.[33]

Finally, taking action that had consequences far beyond what anyone anticipated at the time, in 1970 the DRA staff in Charleston decided to hire a coal miner who could serve as the implicit leader of the movement. They chose a working miner from Cabin Creek who had been one of the instigators of the black lung strike and had continued his insurgent activities in the months that followed. His name was Arnold Miller. One of the founders and chief organizers of DRA summarized the organization's role in reactivating the black lung movement and promoting Miller's leadership:

> What we did was to find the names of other people—a lot of them were from the Yablonski campaign—start communicating with them, and setting up chapters. So, we had the organization of chapters, the lay advocates, the communications network, and the publicity of the Black Lung Association, which meant Arnold Miller. Everything we did for three years was in the name of Arnold Miller. But the whole focus of the thing from our perspective, and for most of the people that were the leaders of the chapters and who were Yablonski people, the whole focus was to change the union.[34]

Although most of the young organizers and several rank-and-file leaders in southern West Virginia viewed the Black Lung Association as a vehicle for union reform, to the miners and widows who flocked to chapter meetings, its self-evident purpose was different: the BLA assisted claimants in obtaining compensation. The divergent expectations and political goals of black lung activists would eventually become a source of division within the movement; for a time, however, the hybrid quality of the association lent it organizational vitality and resilience. The chapters were able to welcome diverse constituents, encompass conflicting beliefs, and meet divergent goals. Their focus adjusted with the changing political context—from black lung compensation to union reform and back to compensation—and they persisted for years beyond what their organizers originally anticipated.

At the center of each black lung chapter were the lay advocates. Trained by the staff of the Appalachian Research and Defense Fund ("Appalred," a public-interest law firm in Charleston with historical origins similar to DRA's), these individuals were able to counsel miners and widows on their black lung claims and represent them during hearings before administrative law judges. Typically, the lay advocates were also the political leaders and organizers of the chapters. Many were committed, long-term activists who had lobbied for the West Virginia compensation law and campaigned for Jock Yablonski, were now organizing black lung chapters, and soon would stump for Arnold Miller and the Miners for Democracy (MFD). These individuals expended immeasurable time and energy contacting miners and widows, arranging for chapter meetings, accompanying claimants to medical facilities for testing, arguing claimants' cases during hearings, and in many other activities. Their work as organizers and claims counselors is probably the single most important reason for the persistence of the black lung movement. To this day, a few of these lay advocates continue to agitate for black lung compensation reform.

The lay advocates were ordinary people by the social and economic yardstick of the coalfields, but they stood out in their leadership and their desire for change. Earl Stafford, for example, was a lay advocate and leader of the black lung chapter in Mingo County, West Virginia. The son and grandson of coal miners, Stafford entered the mines as a boy of fourteen. Vice-president of his local union, active in the Black Lung Association, in the Yablonski campaign, and in the Miners for Democracy, Stafford also ran unsuccessfully for the post of UMWA vice-president for pensioner affairs in the reformed union. Today, he lives in Blackberry City, five miles from his birthplace, on the same creek where four successive generations of his family have lived.[35]

Helen Powell, the daughter and sister of coal miners, is a black woman who became a lay advocate in the Raleigh County Black Lung Association and a leader of the regional BLA. Early in life, she witnessed her father's failing health and disappointing attempts to obtain compensation; by the age of fifteen, she was taking care of him in their home. Powell educated herself about the compensation system in order to assist her father, and she gradually became known in the community as someone who could decipher the bewildering papers and forms that are the currency of bureaucratic institutions. In 1967, she became involved with the Disabled Miners and Widows of Southern West Virginia, and she has been active in different causes ever since.[36]

Most lay advocates were individuals like Earl Stafford and Helen Powell: retired or disabled coal miners, and miners' daughters, wives, and widows. Few had much formal education; most had not finished high school. The training in black lung claims counseling offered them access to professional knowledge that many acquired with diligence and pride. Some became skilled legal reasoners who utilized the law and regulations, the medical arguments of sympathetic physicians, and their own experiential understanding of coal mining to pursue their cases. Their knowledge became a weapon with which to challenge one of the many intimidating institutions—the courts, the Welfare Department, and, in this case, the Social Security Administration—whose elaborate rules, language, and procedures seem designed to mystify and silence, especially their less educated constituents. As one lay advocate stated: "I really like doing that work. I was proud of that, me with only a ninth-grade education. I was proud that I could call up social security and demand what I needed to know."[37]

For many, lay advocacy became like an occupation and provided a similar sense of identification and pride. Few of those who counseled claimants had a paying job or other significant demands on their time; they devoted themselves to the work of the Black Lung Association. Some were older miners who, in a more democratic union, might have risen to district or even international office. Some were women who, in a more open and diversified economy, might have found satisfaction in a full-time job. Lay advocacy offered all of these people an opportunity for service and leadership at a time in their lives when disability, retirement, or the departure of grown-up children from the home threatened to diminish their activity and sense of purpose. For miners who lived each day with debilitation and disease, advocacy was actually a healing experience. Interviewed in 1978, a retired coal miner with black lung commented:

> I liked the work, I enjoyed it. It gave me something to put my mind to. When I'm retired, I can't do no physical work. Being a lay advocate was a lot of work, but it seems like it takes your mind off your sickness. Now I just lie here. I have to take three to six Darvon a day, been taking them since 1957. When I'm here, I don't eat nothing til supper, but when I was on the road, I ate three meals a day. We might—maybe I'll talk to some of the others and just get this Black Lung Association started all over again.[38]

Participants in the Black Lung Association usually came together once or twice a month for a countywide meeting of their chapter. The leaders often opened the meeting by calling on a lay preacher or regular churchgoer to deliver a prayer, which usually included appeals for all to receive their "just due." Individuals invariably took the floor to ask a question about the claims process, make a suggestion, or, most often, testify about the treatment they received from physicians and the Social

Security Administration. Earl Stafford recalled of the Mingo County Black Lung Association:

> We usually had two meetings a month, the second and fourth Saturday or Sunday. A lot of people would bring their papers in, and we'd look them over. I really couldn't tell you how many people we had. There for a long time the building would be full, 200 or 300 at a meeting. Then maybe sometimes there wouldn't be so many, but they'd keep up with what was going on. Lot of people just couldn't come out.[39]

The chapters operated independently of each other. In southern West Virginia, the county boundaries that defined each chapter's constituency are more than arbitrary lines on a map; they represent highly important power structures and the framework of orientation for grass-roots political organizing. For many chapter leaders, their role in the Black Lung Association was a means of achieving respect and recognition in their local communities—a position that many sought to protect. Any coordination across county lines was usually the work of DRA staff members or other organizers, some of whom spent a good deal of time tearing their hair over the chapter leaders' silent refusal to cooperate with one another. BLA leaders attended a meeting in Charleston if they considered it important; they simply failed to appear if they considered it unnecessary. No doubt in some cases they viewed these efforts at coordination as a move to control their chapter or collapse it into a larger organization. One long-time activist and chapter leader, a retired miner, expressed this independent attitude with a single, pithy statement: "There ain't *nobody,* unless I'm doing wrong, gonna tell me how to run my black lung chapter."[40]

Preoccupied with individual claims counseling, members of the BLA spent their days in activities that increasingly resembled those of a service organization. Nevertheless, the movement retained its fiery ideology of class antagonism. Activists' framework for interpreting the apparent indifference and injustice of the Social Security Administration was their experience with the industry that dominated their region and treated them similarly. Physicians who testified in favor of strict eligibility requirements were "company doctors," meaning that they had allegiance to the same interests as the industry. Top officials in social security were likened to the big coal operators, and claims examiners were endowed with the same mentality as company bookkeepers. One angry disabled miner wrote to the *Black Lung Bulletin*:

> Listening to all the miners that have been denied payment for black lung reminds me of slavery days of the 1920s and early thirties. We miners used to load coal for 20 or 25–30 cents a car, work 10 to 19 hours a day, and then when one went to the company payroll office for credit or scrip unless he had his rent, house coal, electric and other overhead company charges covered, he would be refused scrip or credit by the coal company bookkeeper.
>
> Thus it seems the Social Security Administration is loaded down with coal company bookkeepers with the chief bookkeeper in Baltimore, Maryland.[41]

This powerful belief in collective, class-based injustice enabled the black lung movement to unite participants across a chasm of diversity. It counterbalanced the divisive effects of the seemingly arbitrary process by which some claimants were defeated and others enriched. A phrase that became a cliché in the coalfields—"I know a man, never worked a day in the mines, and he's getting his black lung"—expresses this divisive gap which the BLA was for the most part able to bridge.

The movement also united participants across the gulfs of race and gender. The president of the original Black Lung Association, Charles Brooks, was a black miner from

Kanawha County, and in subsequent years several of the most visible leaders and lay advocates were black women and men. Racism existed, of course, but the strong belief in collective injustice and entitlement mandated against its expression. One of the common jokes of the movement, told in various forms by both black and white, reflects this unity born of common hardship: "I used to kid with these black men, some I know pretty well. Some could get paid, some couldn't. Well, I said to them, 'You all were *born* with black lung, and they won't even pay you.' "[42] Racial unity also had deep historical roots in the workplace, in the United Mine Workers, and in the coal camp community. One white miner, a lay advocate, stated: "I was raised with colored people, worked with them, too. . . . We'd go up to Washington together, stay together in the same room, eat together. No, I don't believe in that discrimination."[43] Speaking of the Black Lung Association, one black miner commented: "I never saw racial discrimination surface in any manner. It's one of the few things in my life where I never saw it."[44]

The role of women in the Black Lung Association varied according to changes in the movement and political developments in the United Mine Workers. The original Black Lung Association was composed entirely of male coal miners, but women soon became involved in the movement on their own or through their husbands. Many women went through training to become lay advocates; they often had better reading and writing skills than men, having stayed in school rather than going in the mines at a young age. A few women emerged as regional leaders of the Black Lung Association, but in the local chapters, women often worked behind the scenes. Some functioned as organizers, those who did the maintenance and logistical work necessary to hold a chapter together, but they stood in the background as men conducted meetings and delivered speeches. With the victory of the Miners for Democracy in 1972, BLA leaders began jockeying for access to the resources of the UMWA, placing women at a further disadvantage, because they were not union members. Coupled with the gradual decline of the association, beginning in 1973, this problem led some women to continue lay advocacy on their own, without formal ties to a black lung chapter, or, in one case, to set up their own organization, the Association of Miners' Wives and Widows.

Whatever its internal tensions or weaknesses, the Black Lung Association appeared to the Social Security Administration as a phalanx of angry constituents. In addition to offering the popular service of claims counseling, by 1971 the BLA chapters were also maintaining a visible, embarrassing campaign of direct pressure on agency officials. With a shrewd eye for the media value of their actions, black lung activists staged demonstrations and set up informational picket lines in front of the local offices of the Social Security Administration. They raised demands that reflected not only a desire for more equitable eligibility standards, but also an insistence on better treatment by agency employees. "We demand Social Security treat us like people, not like dogs," declared the leaflet that BLA activists in southern West Virginia handed out daily at the entrance to the Logan social security office.[45] The protestors demanded that the agency process claims more rapidly ("Don't wait until Black Lung kills US"), collect all pertinent evidence in each case, and inform claimants about the precise reasons for their denial.[46] "Claimants' rights"—to be examined by physicians of their choice, to review their own claim file, and to receive copies of the file's contents—were prominent in the leaflets and verbal demands. The demonstrations, picket lines, and local media attention publicized claimants' grievances with the Social Security Administration; they eventually yielded more open policies in several local offices and provoked a response from top officials at agency headquarters in Baltimore, who agreed to come to the coalfields and hear out the complaints.

By the spring of 1971, the BLA chapters, which now stretched into eastern Kentucky and southwestern Virginia, had coalesced around a legislative agenda of federal black lung compensation reforms. The groups supported a bill introduced into the U.S. House of Representatives by Congressman Carl Perkins (D-Ky.) which would prohibit the use of X-ray evidence as the sole basis for denying a claim, eliminate the offsetting of benefits between social security disability and black lung compensation, extend federal funding of the program for two years, and permit payment of benefits to double orphans.[47] In addition, the Black Lung Association sought the following:

> We demand that all eligible miners and widows have a right to complete and impartial examination.
>
> We want properly equipped clinics, to give tests to miners, throughout the coal fields.
>
> We want proper assistance in filing and processing claims.
>
> We want non-medical evidence accepted in Widow's cases where medical evidence is not available.[48]

The campaign for what became known as the "'72 amendments" was, in the words of one organizer, "the most effective grass-roots lobbying [he had] ever seen."[49] Over the next year, activists traveled in car caravans to Washington, D.C., lobbied their congressional delegations, and periodically picketed the local offices of the Social Security Administration in a united effort to liberalize Title IV of the U.S. Coal Mine Health and Safety Act. Although the wildcat strike over black lung compensation in West Virginia was never repeated on the same scale, the strike potential always loomed as an implicit threat behind BLA demands. In one instance, the leadership ended a formal statement to the Social Security Administration with the warning: "We have been restraining our people as much as we can.

We will not continue to do so within two weeks."[50]

In the forefront of the campaign were the lay advocates, who, because of their training and claims experience, thoroughly understood the legal changes that the movement sought. Some reinforced their demands regarding technical points with more earthy arguments:

> We'd go to Washington in a group and just go from one office to another and not sign the guest book. You know, when you go to Washington, they don't see you as a person, they see you as a vote. It's a little bit trickier with the ones not from these coal states. One senator, he said, "Why should my state compensate these miners and we don't have a coal mine in the whole state?" My husband, he said, he just put it like: "Because it's the coal miners that keeps your ass warm, that's why." To me, there's nothing or nobody that stands higher than the coal miner.[51]

On Capitol Hill, curious alliances and complex negotiations accompanied the legislative reforms as they wended their way through subcommittees, floor debates, and a conference committee toward final congressional approval. All segments of the coal industry supported extension of federal financing for black lung benefits, but the provision provoked an outcry from liberals and conservatives alike, who for different reasons were eager to push the cost of compensation onto the operators. Large companies with heavy investments in underground mines also had no quarrel with proposals to extend benefits eligibility to surface miners; this would force strip mining companies to share in the liability for black lung compensation whenever federal financing was terminated.

As usual, the legislative provisions pertaining to the definition of disease and disability kindled the most heated controversy. Taken as a whole, the most liberal proposals legitimated in technical form the black lung movement's experiential concept of disease. If embodied in the law, they would help to make a generic "black lung," as opposed to a single coal workers' pneumoconiosis, the basis of compensation eligibility. The legislation repudiated the use of X-rays as the sole basis for claims denial and mandated consideration of "all relevant evidence." Blood gas studies, electrocardiograms, and other diagnostic tests would all be taken into account, regardless of X-ray results. Moreover, lay evidence—the nonprofessional observations of co-workers or others acquainted with the miner's physical condition—could be submitted as legitimate support for a claim. The proposed amendments granted miners with fifteen years of underground (or comparable) employment and evidence of totally disabling respiratory disease a rebuttable presumption of total disability due to pneumoconiosis, even in the absence of a positive chest X-ray. The definition of disability was relaxed by tying the definition to a miner's ability to perform his customary job in the mines. (Previously, the Social Security Administration had argued that claimants were not totally disabled if they could perform a light, sedentary job—few of which were actually available in the coalfields.) Finally the legislation required a review of all pending and previously denied claims under the new eligibility standards.[52]

To physicians who believed that CWP was the only disabling work-related respiratory disease of coal miners, the legislative proposals were illogical, even absurd. Presuming a miner "totally disabled due to pneumoconiosis" when he had no legitimate scientific—that is, X-ray—evidence of the disease contradicted their professional practices and integrity. Physicians, coal operators, officials of the Social Security Administration, Republicans, southern Democrats—all lined up to defeat the legislation. They failed. On May 19, the Black Lung Benefits Act of 1972 was signed into law.

The legislation passed despite formidable opposition, in part because key congressional supporters were well positioned to push it through.[53] Congressman Carl Perkins was chairman of the House Committee on Education and Labor, and John Dent of Pennsylvania headed the House Subcommittee on Labor. Due to BLA organizing, many of Perkins's constituents were well aware that Kentucky had the lowest black lung claims approval rate of any state in the country, and in this election year, their discontented voices were amplified. In fact, the original "Perkins's amendments" also received some support from within the industry, but these amendments were far more modest than the legislation that eventually was passed into law. Once a black lung bill was introduced, liberal Democrats from non-coal-producing states promoted several of the most far-reaching reforms, as they had for the 1969 Coal Mine Health and Safety Act. They had little to lose by crossing the coal industry, and something to gain from their own labor backers by responding to coal miners' grievances and supporting a federal compensation program. Politicians like Phil Burton of California had a more personal account to settle: those who had worked hard to draft and promote the original black lung compensation legislation were exasperated at what seemed to be agency contravention of congressional intent. They were determined to force social security to compensate all miners who were disabled by occupational respiratory disease. The coal industry itself may have contributed inadvertently to the legislation's success by pushing hard for an extension of federal responsibility for black lung benefits. The annual cost of the program was already five times greater than the largest original estimate; few legislators of any political stripe, except those from coal-producing states, were in any mood to continue subsidizing the cost of one industry's occupational disease. Finally, the legislative debate over black lung reform once again was conducted against a backdrop of political turbulence and aggressive rank-and-file action within the United Mine Workers. In 1972, the Miners for Democracy were making labor history as they gathered for a nationwide convention in Wheeling, West Virginia, and laid plans to regain control of their union.

The 1972 amendments were a clear-cut political victory for the black lung movement—much more so than the West Virginia compensation law for which the movement is known. More politically sophisticated and well organized than in 1969, BLA leaders understood that the substance of their accomplishment could vanish if the Social Security Administration promulgated stringent standards regulating the program. They therefore initiated an intensive campaign to force the agency to permit their participation in composing the new regulations. BLA leaders traveled to Washington, D.C., and made their demand in person. Pickets appeared at local agency offices in West Virginia, Kentucky, and Virginia. Pressure finally resulted in an important and unusual concession: the Social Security Administration agreed to allow a committee of BLA representatives to come to Washington at government expense and participate in drawing up the regulations that would govern administration of the black lung compensation program. As a result, the Black Lung Association won extraordinarily liberal eligibility standards for all claims filed before July 1, 1973; however, claims filed after that date would be administered by the Department of Labor and judged under more stringent regulations.[54]

One lawyer closely involved with the black lung movement recalled:

> "When the 1972 amendments were passed, everybody was waiting. For a while, there wasn't any more demand for help. Plus, Social Security was writing the regs that finally came out in September. During that summer, there was this time when black lung people went to Washington and had this knock-down, drag-out session with Social Security. That resulted in the interim standards, under which many, many people were paid. What I always thought about that was that the standards were outstandingly liberal in terms of

> people who were already disabled, but they let the permanent standards go, so that after July of 1973, practically no new claimants got paid. That kind of sold out the working miners for the ones already disabled, which I think probably reflected the needs of the people doing the negotiating. Both Social Security and black lung people referred to that later and said it was the only time in history that Social Security sat down and really negotiated with the black lung people. People who were involved in that—it made them feel very powerful.[55]

With passage of the 1972 amendments, it appeared that the black lung movement had come to an end. Reform of the statute and promulgation of the interim standards seemed to satisfy the Black Lung Association's goal: claimants now would be treated fairly. Former BLA leader Arnold Miller was running for president of the UMWA on the slate of the Miners for Democracy, and many chapter activists shifted their energy into the election campaign. DRA disappeared from sight, its organizers absorbed into other causes. Remarkably, however, the black lung movement once again reemerged. During its final phase, many of the weaknesses inherent in a single-issue movement for compensation became apparent. There was a period of brief renaissance followed by bitter feuding and, eventually, decline. It was a period when the battle for black lung compensation became increasingly caught up with the movement for union reform, as participants in both efforts wrestled with the divisive and troubling aftermaths of their own victories.

# 6

## Black Lung and the Politics of Union Reform

Progressives all over the country cheered on the officers and campaign staff of the Miners for Democracy as that organization took over the United Mine Workers' headquarters in December 1972. The miners' victory was historic: they had built the first rank-and-file insurgency in the United States that won formal power in a large, industrial union. Perhaps the labor movement was on the eve of a new era, when the vigor of rank-and-file leadership would be restored.

As the elation of victory began to diminish, however, the appalling complexity of the tasks the insurgents faced became clear. By winning institutional power, the reformers had surmounted only the first barrier in an obstacle course of administrative chaos, political intrigue, and internal dissension. The insurgent leaders had pledged to discard many of the union's policies and to democratize its autocratic structure; in practice, this meant virtually reconstructing from the ground up a large and complicated institution. Miller's report to the first UMWA convention after he assumed leadership of the union expressed his frustration with the administrative aspects of union reform:

> I think it should be pointed out that we underestimated the extent to which the central administrative machinery of the UMWA had just plain stopped functioning. It had the *appearance* of functioning,

> but when we blew the dust off and turned the switch, the whole apparatus fell apart at our feet.[1]

The reformers faced complex strategic questions, including how to reconcile the principles and procedures of democracy with the political need to subdue the Boyle machine, which continued to thrive in several districts. Also looming was the question of how the new union officials should relate to their own rambunctious constituency, the insurgent miners who had elected them. Finally, there was the coal industry: reformers had promised many changes that required success at the bargaining table. How the industry would respond to the new union demands was entirely unclear; the operators were edgy, apprehensive about what this rank-and-file takeover would mean.

For miners and their families, 1972 was a victorious year, but the tenor of the years that followed was more ambiguous. The movements for black lung compensation and union reform, although increasingly divergent in their goals and social bases, faced similar political dilemmas. Both had won significant victories that, in the eyes of many constituents, signaled the end of their struggles, the achievement of their goals. Yet, their singular triumphs—passage of the 1972 amendments for the BLA, and the election victory for the Miners for Democracy—did not in and of themselves guarantee lasting change. In the years after 1972, these movements were challenged to institutionalize and defend the reforms they had sought, yet both were hard pressed to muster the leadership and the organized base of power required to do so.

Convoluted political struggles took place within both movements in the aftermath of victory. Although in some cases straightforward contests for power, the struggles also centered on differences in political vision and strategy. Was the ultimate goal of the black lung movement a liberal compensation program, or prevention of disease in the mines? Was the ultimate goal of the Miners for Democracy the ouster of Tony Boyle, formal revisions in the UMWA constitution, or fundamental changes in industrial relations with the coal operators? Basic political differences that had seemed unimportant when confronting a common enemy now came to the fore. In the years after 1972, those who sought far-reaching changes in the workplace and in the union attempted to rechart the course of these movements. Their task was enormous and their chance of success slight. The political framework of each movement had been forged years earlier. Those who now attempted to reconstruct it began their project quite late, perhaps too late.

## The Anatomy of Decline

> In 1973, the Mingo County Black Lung Association was probably in its most dynamic period, at its height. Right after the '72 amendments, a whole lot of claims were in fact getting paid. Through '72 and on into '73, they were getting large numbers of people who would come out to meetings, a small core of whom were doing claims.
>
> In Logan, there wasn't much of an operation. . . . During the summer of '74, we went to organization meetings, did some counseling. I remember Willie Anderson came to one meeting. He stood up and said something and asked some questions, and it was obvious that he had some knowledge about what was going on. What ended up happening that summer, some new people came forward—Martin Amburgey, Willie Anderson. The organization got pulled together again.[2]

The leadership and vitality of the Black Lung Association chapters in West Virginia shifted considerably after 1972. Chapters in which key leaders went to work for the union, such as that serving Kanawha and Fayette counties, collapsed. Other groups, especially those with energetic, indigenous leadership, capable lay advocates, and organizers willing to do legwork, enjoyed a resurgence. Chapters thrived not only in Logan and Mingo counties, but also in Raleigh County, where activists put together probably the largest black lung organization in the state.

The revival of the Black Lung Association in West Virginia was due in part to the liberalization of the federal compensation program during 1972. Under amendments passed in that year, all previously denied claims—some 174,000—were subject to review under more lenient eligibility standards.[3] The review of claims generated a sudden, heavy demand for the services of lay advocates, while the relaxed eligibility standards guaranteed them a high success rate. These developments boosted the prestige and popularity of the BLA chapters, especially those that could offer miners and widows assistance with their compensation claims.

Behind this resurgence, however, lurked serious weaknesses that would soon become evident. When it originated in 1968, the black lung movement drew leadership and participation from retired and disabled miners, widows, and, above all, older workers who were still in the mines. Their common grievances ranged from black lung compensation to union reform, from pensions to workplace health and safety. By 1973–74, however, many of these older workers had left the labor force, and the vital, direct link between working miners and the black lung movement was severed. This diminished the movement's essential economic leverage and base of power, which ultimately resided with working miners. New workers, of course, entered the mines, but for this younger generation, the struggle within the union and the workplace pressed with greater urgency than the more remote problems of chronic lung disease and disability compensation. The black lung movement increasingly found its followers among retired miners and widows who had a single goal in mind: winning their compensation claims. Individuals who succeeded often dropped out of the movement. Thus the renaissance that followed the 1972 amendments was temporary. By the end of 1974, over 350,000 claims for federal black lung compensation had been approved, and the social base of the movement shrank accordingly.[4]

Activists' success in revising the eligibility criteria for compensation brought other ironic consequences. By 1974, mastering the intricate rules governing federal black lung compensation was an exacting task even for experienced lay advocates. The original statute authorizing the program had been modified by amendments in 1972; the original regulations governing the claims process had been displaced by the "interim standards," which in turn were superseded for certain claims by permanent eligibility standards; the original administrative agency for the program, the Social Security Administration, now handled only certain claims, while the Department of Labor processed others utilizing different procedures and eligibility criteria; and, finally, many disputed claims had been settled in the federal courts, so that various judicial decisions and legal precedents also entered into the determination of eligibility. A complex program to begin with, given the controversial and ambiguous medical issues involved, black lung compensation had become Byzantine even by the standards of those accustomed to unraveling tangled legal principles. In an opinion rendered in 1976, one federal judge complained, "The statutes and regulations governing pneumoconiosis determinations are almost unfathomable."[5]

Complexity was not the only problem with the program. Despite five years of struggle, the political institutions that had yielded to the pressures of the black lung movement obstinately resisted permanent implementation of certain reforms. The Department of Labor seemed to have a penchant for restrictive interpretations of the statute. Although many claimants were awarded benefits under the interim regulations, others whose disability seemed equally severe were turned down under the permanent regulations. By 1974, the denial rate began to climb.

This trend demoralized many of the chapter leaders and lay

advocates, who had become increasingly preoccupied with the intricate, changing details of the claims process. One organizer reflected:

> Participation was all based on the ups and downs of the claims thing. They agreed to let some older people through in the '72 amendments in exchange for a drastic cutoff. By 1974, it was pretty clear that they were really slamming the door in people's faces. Rereading of X-rays began to be much heavier. The number of turndowns was so high, demoralization was creeping in. There was more infighting between the chapters. Many of the leaders were getting tired, and some were getting cynical, grumbling about how people don't support you when you help them.[6]

As ever, most of the claims denials involved the relentless, underlying controversy over the medical definition of black lung. By law, a negative chest X-ray could no longer be the sole basis for denying a claim; however, under social security's liberalized regulations, an X-ray revealing simple CWP (plus ten years in the mines) could establish eligibility. Thus, chest X-rays remained critical to the claims process, and battles over their interpretation persisted. The Social Security Administration continued to utilize so-called "B-readers"—that is, physicians who had passed the agency's examination on X-ray interpretation—to reinterpret claimants' chest X-rays. Critics asserted that the examination simply reflected the agency's conservative viewpoint: physicians who tended to classify X-rays as positive for pneumoconiosis failed the test; those who did not as frequently "see" pneumoconiosis passed.[7] The B-readers did tend to reclassify X-rays "down," indicating that they revealed a lesser stage of CWP than the X-rays' original readers had diagnosed. One sample of 164 approved claims reviewed by the U.S. General Accounting Office revealed that B-readers had reclassified positive X-rays as negative in 35 percent of the claims; in only 10 percent had they reclassified a negative X-ray as positive for CWP.[8]

A routine procedure on many claims, the rereading of X-rays reached absurd proportions on a few occasions. Consider the following summary of a black lung claim that eventually wound up in federal court:

> [Nineteen seventy] and 1971 x-rays read negative. July 1973 x-ray initially read positive; re-read positive at a.l.j.'s [administrative law judge's] request; then re-read negative at a.l.j.'s request by a B reader. February 1974 x-ray read positive by a B reader; re-read positive at a.l.j.'s request; re-read again at a.l.j.'s request by a B reader. May 1974 x-ray not initially classified for pneumoconiosis; re-read positive at a.l.j.'s request; then re-read negative by a B reader. A.l.j. concluded x-rays were negative.[9]

The court remanded this case for payment of benefits, observing: "Certainly, if the Law Judge continues to seek re-readings *ad infinitum*, he will eventually come upon a conservative reader who will find the films to be negative. However, in the process, the Law Judge renders the non-adversary adjudication of claims a meaningless exercise."[10]

The criteria for assessing disability also remained problematic. Some physicians charged that the numerical scales utilized to quantify "total disability" were inconsistent: the scale applied to the results of blood gas tests was unjustifiably stringent, far more so than the one used for pulmonary function tests.[11] Moreover, although the liberal interim regulations greatly relaxed the standard for total disability as measured through pulmonary function studies, the permanent regulations tightened the standard; under both sets of criteria, test results were often reevaluated and disqualified by medical consultants. That happened, for example, with the following black lung claim from southern Ohio: "An October 1974 report by an internist included a positive x-ray and qualifying pulmonary function test. Social Security consultants re-read the x-ray negative and the pulmonary function test as showing inadequate effort."[12] The miner appealed his claim all the way to federal court, which remanded the case back to the agency.

Although now eligible for black lung benefits if her husband was disabled by legally defined "pneumoconiosis" at the time of death, a widow still encountered special obstacles in the claims process. Those whose husbands were still working in the mines when they died were automatically denied compensation; according to the agency, such employment constituted prima facie evidence that the miners were not "totally disabled." This argument was used inflexibly, even when affidavits indicated that a miner worked only under economic duress and with substantial assistance from co-workers. Consider the following case from western Pennsylvania: "Miner had worked 34 years in mines. Miner's doctor had advised him to quit because of pneumoconiosis, and he had intended to, but he was killed [in a mine accident] before he could carry out his intention."[13] The widow's claim was denied.

Claimants' difficulties in establishing eligibility grew far worse under the Department of Labor's administration of the black lung program. The 1969 Coal Mine Health and Safety Act assigned to the secretary of labor responsibility for overseeing the transfer of the federal black lung program to the states' workers' compensation system. Because no state offered a comparable program, however, black lung compensation remained under the aegis of the federal government. Claims filed after July 1, 1973, were processed by the Labor Department.[14] Miners and widows who sought compensation through the new administering agency did not retain the benefit of the liberal "interim standards"; their claims were judged under permanent and more exclusive eligibility criteria. With the stroke of a pen, officials in the Department of Labor revised the medical evidence requirements for compensation and redefined the meaning of total disability. A chest X-ray showing simple pneumoconiosis was no longer sufficient to establish eligibility. Miners also had to meet far lower numerical standards—their breathing capacity had to be far more impaired—in order to qualify as totally disabled. Black lung compensation was thus placed out of reach for most new claimants: the department turned down 93 percent of all claims.[15]

Beginning on January 1, 1974, liability for new claims shifted from the federal government to the coal industry. Due process required that a company be allowed to contest any claim for which it might be held liable; as a result, the claims process became adversarial: coal company lawyers could challenge claimants' efforts to obtain compensation. The industry aggressively fought liability for black lung benefits, not only by contesting 97 percent of all claims but also by challenging the constitutionality of the program in federal court. Resistance from industry, coupled with the restrictive eligibility standards, resulted in lengthy delays and a high denial rate for claimants. After three years of industry liability (1974–76), the operators were funding a grand total of 123 black lung claims.[16]

The Labor Department's inaccessibility compounded claimants' exasperation with the program and shielded the department's administrators from claimants' influence. In the local social security offices, where, prior to 1973, miners and widows went to file claims, the agency staff was accessible; furthermore the offices provided a place where activists could educate claimants and build their organization through picket lines and leaflets. The U.S. Department of Labor maintained no comparable network of local offices; it relied almost entirely on written correspondence with claimants. The agency did not offer them a toll-free information number—indeed, did not include any telephone number on letters to claimants. Added to the denials, delays, and bewildering regulations, the department's inaccessibility produced enormous frustration among many claimants, for whom there was no longer a local target at which to direct their discontent.

Perhaps the single most tragic result of this situation was the murder of one dedicated lay advocate and miner's widow, Anise Floyd, by a desperate black lung claimant she was representing. Her close friend and fellow lay advocate, Helen Powell, wrote a eulogy to Anise Floyd in which she stated:

> Her death came at the hands of a man who was her friend, who she had helped. Sick and deranged though his actions were, we cannot dismiss his actions as simply those of a lunatic. This man has worked all his life and had little to show for it except a crumbling body and a frustrated spirit. He had been waiting month after month for more than two years for any work whatsoever on his Department of Labor black lung claim.
>
> We who work with claimants know all too well the frustration that develops watching the mailbox day after day, week after week, for word about a decision that can mean all the difference between a little comfort at the end of life and continuing poverty.
>
> We know the rage that develops in people whose fate will be decided by laws and regulations so complex and arbitrary they have no way of comprehending them.[17]

By 1974–75, the recognition that their movement was falling apart did not escape many Black Lung Association leaders, lay advocates, or organizers. Signs of decline were abundant. Attendance at county chapter meetings fell off. Denials for federal black lung compensation greatly outnumbered approvals. Political activities now consisted mainly of legislative lobbying, with the same small groups of people occasionally traveling to Washington, D.C. Some saw in this decline the potential—indeed, the requirement—for a change in the strategy of the black lung movement. Those with a stake in the future fairness of the compensation program—working miners—were growing in number and militancy. BLA activists with a historical perspective knew that the strike action of coal miners had always been the movement's ultimate source of power; now perhaps working miners could stimulate the movement's renewal. One organizer recalled:

> During the summer of 1974, there was a black lung meeting at Cedar Grove. Some of it was around what the Black Lung Association was going to do in the future. To the degree that there was an orientation that was adopted, it was that you had to get working miners into the movement. It had become clear that they would not get their benefits. So it was adopted and carried out to some degree—to go out to locals and talk about health, the dust control program, black lung. Underneath it all was the thrust to organize a strike on black lung reform.[18]

This decision came at a time when one central political issue was emerging with increasing clarity in the coalfields, especially in southern West Virginia: the inability of the reform administration to redefine the role of the union and to restructure industrial relations in a manner that could satisfy the most radical segment of the rank and file. During 1975–76, coal miners increasingly took these tasks upon themselves, through one of the most extraordinarily militant and protracted wildcat strike movements in U.S. labor history. Black lung activists sought to revitalize their movement by placing black lung compensation on the agenda of this rank-and-file upsurge. The prospects, however, were not favorable; their numbers and vitality had ebbed to a dangerously low level. Moreover, since 1972, the contentious interaction between activists and the new officials of the United Mine Workers had served only to hasten the decline of the black lung movement.

## Between a Rock and a Hard Place

Fissures in the rank-and-file base that brought the Miners for Democracy to power began to appear not long after the 1972 election. The hazardous work of unifying insurgents around opposition to Tony Boyle and support for key issues like autonomy soon began to seem easy compared to the torturous complexity of making reform an institutional reality. Disputes among leaders and staff impeded the formulation of coherent policies to guide the union, and administrative inexperience compounded the episodes of confusion. Despite the MFD election victory, even political control of the union was not secure; key district offices and seats on the International Executive Board (IEB), chief policy-making body of the union, were captured by MFD opponents.

One of the first critical political decisions the reformers made was to dismantle their own insurgent organization, the Miners for Democracy. Most believed that "the political party, MFD, basically had *become* the institutional union."[19] Few saw the need to continue it, and some were no doubt fearful of the alternative locus of power that it might represent. Keeping MFD alive would have required deliberate and extensive organizing, as it was always a skeletal network of rank-and-file leaders. With the election victory, many activists joined the union's staff or became preoccupied with running for office in their now autonomous districts. The decision to disband MFD had serious consequences, however: it left the new administration without a coherent rank-and-file base, and it left the rank and file without an organized vehicle to force the accountability of their leaders. Reflecting on the decision to dismantle the Miners for Democracy, one former UMWA staff person commented:

> That was the single greatest mistake that was made in the whole effort—in my opinion. Now there're a lot of people who disagree with me about that. I came to Washington with Arnold's victory, and this was one of the first big policy issues that had to be discussed. And there were two schools of thought. One was that it's time to make peace. If you have MFD, then you've got a faction that's going to keep the union split. Another school of thought, to which I subscribed, was that the battle was far from over, that at least insofar

> as the hard core that had won that election—that's miners, folks that had gone out and worked for Arnold hard—were concerned, it was really an ideological question. It wasn't a matter of replacing one face with another; there needed to be an ongoing vehicle for generating and disseminating the types of views and building the types of organizations that would secure the victory that Arnold had won. There was a great deal of discussion and argument over that.[20]

Once in power, reformers moved quickly to democratize the union's internal structure and to arrange for elections in various districts. Believing that the democratic principles on which they had run required a policy of neutrality, the top union officials did not become involved in most of these district elections and refused to campaign for reform candidates. The result was a further weakening of MFD's political base. Key districts went to the old machine, and reformers soon lost control of the International Executive Board. In the view of one staff person: "What ultimately happened to [Miller] was he lost. He won the presidency, lost the Executive Board, and lost his power to a greater or lesser extent."[21]

Even in areas where they exercised formal authority, such as the union's top administrative functions, reformers were unable to keep their ranks together, make decisions, and move forward. The antagonism between Miller and Mike Trbovich, UMWA vice-president, did not abate but worsened, and rivalries among other employees and departments soon developed. Miller was unable to unite and lead his own staff, much less pull the union out of the administrative and financial morass created by decades of cronyism and corruption. His own lack of experience and expertise was compounded by the inexperience of his staff, one of whom said:

> A lot of us, and I hold myself responsible as well, for the first time had to work in a real organization—with a social structure, a political structure, with a history. It was the first *real* job many of the people had. Secondly, a lot of the people who came in with Miller

were "outsider" types, who saw the union as a base for pressing their particular personal interests. So the union would end up with four positions on strip mining on the Hill, because one person's position differed from another's, which differed from a third person's, and they were all up there talking their thing.

The third factor, and I can appreciate it and understand it, was that there was a concern that everything that had been there was bad. Everybody who was associated with or who was on the payroll of [the union] was bad. So everything, everybody, was thrown out. There was no bridge from the old to the new. And I can understand that, but I don't think it was exactly true.

There was never a manager. You're running a multi-million-dollar operation, and there's no manager of it. It sort of became shoot-out at the OK corral—staff fighting among each other, Arnold playing one off against the other, playing favorites this week and switching back and forth. And I didn't contribute very much to changing that.

I wouldn't trade that experience for anything. For most people, it was the first time they had been involved in anything that they won. They didn't know what the hell to do with it. The potential, the commitment, all those things were there. People really gave a shit, but there was a lot of naiveté, a total disregard for anything that had been there before, a total disregard of history.[22]

Ignorance and naiveté had origins in a history that both produced and thwarted the Miners for Democracy: five decades of dictatorship generated a longing for democracy, but they also robbed miners of the skills and experience required to run their own union. Miners had been denied not only the opportunity to exercise leadership, but also the experience of active, democratic participation in the union's internal life:

There was a conscious effort on the part of John L. Lewis and later Boyle *not* to have people have the tools that would make them aggressive trade union activists, who had acquired certain skills, certain experience, certain political involvements; they were afraid of that. There was no opportunity for a bright, aggressive young coal

> miner to, through his union, gain experience; you couldn't move on up, you couldn't take the next step, because all those jobs were appointed out of Washington by Boyle.[23]

The legacy of this era was not simply inexperience on the part of Miller and other MFD officials; especially under Lewis's forceful and charismatic leadership, the post of union president had acquired mythic proportions. Many miners retained highly inflated expectations concerning what the union president could and should do. MFD's failure to organize extensively among the rank and file played into this legacy:

> If I would view a major mistake of the movement, it was that there never was the grass-roots support formed. The legal arguments against Boyle just kept building and building. So what happened, from my perspective, anyway, was they could have put Howdy Doody up there and won. There became a kind of groundswell to just push Boyle aside and get a whole new group up.
>
> Plus there is a kind of religion about the union which made the president the god. The kind of letters I remember Miller receiving, calls and what have you, that the only way they could solve their problems was for the president to solve them, because of the history of John L. Lewis, the history of that union, the history of top-downism. The Man will take care of me, so there's no need to have an organizational base.[24]

Thus, the insurgents-turned-officials were caught in a political vise: on one side stood the rank and file, who expected swift resolution to a multitude of institutional problems and individual grievances; on the other side were MFD's opponents, who gained control of the International Executive Board and proceeded to blockade implementation of the reform program. IEB meetings became scenes of "protracted and agonizing *total war*," in the words of one staff person.[25] The discord and tension took their toll. Miller, the new president, developed a reputation for disappearing whenever tough decisions or heated confrontations were in the offing. Staff members became disillusioned and quit, one by one. The administrative chaos increased. Just keeping the doors open and the lights on at UMWA headquarters became an accomplishment.[26]

These developments reverberated through the black lung movement. Demoralized by the small attendance at meetings and the overall reduction in their social base, undercut by regulatory changes in the compensation program, activists now found themselves on opposite sides of the battlefield of union politics. Splits appeared within the already weakened movement. Miller supporters were suspect in the eyes of those who had become disaffected. Resentment simmered between leaders who received financial support from the union and those who did not. Rumors and squabbles accompanied the factional trend, and personal insults often obscured important political differences that were also at issue.

The election victory of the Miners for Democracy reshaped the political landscape in which black lung activists pursued their goals, yet the movement as a whole never adapted to the new terrain. Miners had founded an organization distinct from their union because UMWA officials were uniformly unresponsive to their demands. Now that a rank-and-file insurgent and former black lung leader was at the helm of the union, the continuing role of the Black Lung Association was in doubt. What should be the forum for articulating rank-and-file goals concerning black lung: the UMWA or the BLA? What should be the stance of the Black Lung Association toward the new union leadership: loyal defense, critical support, or opposition? To what extent should the Black Lung Association rely on the institutional resources, including financial, of the union? These were the troubling questions on which BLA leaders failed to develop a coherent position that might have unified their movement and guided their relationship with officials in the UMWA.

At the ends of a political spectrum that included many intermediate and contradictory positions, there developed two opposing perspectives on the BLA's relationship with the new administrators of the union.[27] Miller drew his firmest support from miners who had been his companions and fellow leaders since the movement's early days. They were members of the same generation who had spent their working lives in the dusty mines of the postwar years and then had stood shoulder to shoulder in the black lung strike, the Yablonski campaign, and the Miners for Democracy. Most of these individuals viewed the reform administration through the prism of their personal friendship with Arnold Miller. Many eventually became disillusioned: some did not enjoy the personal access to Miller which they felt was their due; others became angry over the dual pension system introduced in the 1974 contract, which created a wide differential between early and more recent retirees. Disaffection occurred at varying speeds and for different reasons, however, and eventually fragmented rather than unified these members of the "old guard."

Others in the black lung movement were from the start more skeptical of the Miller administration. These individuals tended to identify with the more militant segment of the rank and file (rather than with the elected leadership), and they sought reforms that far exceeded an exchange in personalities at the top of the union. Some interpreted the dissolution of the Miners for Democracy as a power play executed to minimize the new leaders' accountability to the rank and file. They feared similar attempts to disband the BLA: "Miller and them wanted the Black Lung Association destroyed. They didn't want no dissidents, you know, people speaking out."[28] The reform administration's failure to proceed rapidly with implementation of the MFD platform seemed to confirm these skeptics' suspicions of betrayal. Those within the black lung movement who expressed this attitude included miners who had not identified strongly with MFD to begin with, as well as some non-UMWA members (e.g., widows, certain disabled miners) who had little access to the union. Complicating this scenario was the presence of a new crop of young organizers, several of whom hailed from rival organizations on the Left and imported their own disputes into the black lung movement. Like the Miller loyalists, members of this opposition group failed to cohere around a single, consistent position regarding the union leadership.

As activists began to compete for access to the union's resources—its finances and institutional clout—the fractures worsened. In a movement that had long welcomed diverse constituents, a single attribute, UMWA membership, now elevated some above others. The seriousness of the division between those who were in favor with the union and those who were not became unmistakably apparent at the UMWA Pittsburgh convention in December 1973, the first for the reform administration. One union staffer recalled:

> The Pittsburgh convention was two weeks long and very intense. Some of the Regional BLA people were traveling together and came to the convention, but none of them were union members, and none could vote or have an input compared to what BLA people who were in the union could. The union people were all on committees; they were an integral part of the decision-making process. We had managed to get them on one or another committee, so that BLA people who were in the union were well represented.
>
> I think the convention was a watershed in terms of splits between the union faction of the BLA and other people. Because after the convention, there was the [UMWA] Legislative Department always in the black lung business, and also that convention was a very concrete thing that sort of defined people as in or out. It wasn't like other meetings where people could just come.[29]

Financial assistance from the union became another bone of contention. Declining chapter membership increased the need for outside support to pay for a local office and lobbying trips to Washington, D.C., as well as to meet other organizational expenses. Some activists wearily remember this period as one of incessant discussions about money—who was getting it and who was not. One lay advocate recalled:

> During that time, there was a lot of feeling around money, and it was a source of contention between people. Some people didn't do anything, didn't do claims, and they would get money. Others really put themselves out as lay advocates but went through terrible times where, when they came to Washington, they didn't have money to eat or anywhere to stay. After the union people got money, they stayed at nicer places like the Pick and even the Hilton, and the other black lung people stayed at the Herrington, which is clean but seedy.[30]

Precariously situated between the black lung movement in the coalfields and the reform administration in Washington, D.C., was the Field Service Office, an outpost of the union that Miller established in Charleston, West Virginia. This office became an important avenue of interaction between the Black Lung Association and the United Mine Workers. One of its staff members recalled: "The first idea of the Field Service Office—it was a real strong idea then—it was Arnold's office, and it *was* his office at first. That office was to be his base in the coalfields."[31] However, as Miller retreated from the demands of black lung activists and rank-and-file miners, he gradually abandoned the Field Service Office and its staff: "People originally had a lot of loyalty to Arnold; they thought they were friends and would have access to him. It turned out both organizationally and individually none of those people had any access to him."[32]

Under the leadership of its first director, Linda Fritts, the Field Service Office assumed several functions, such as coordination, communication, and claims service, which Designs for Rural Action had provided the black lung movement in an earlier period. Many members of the initial staff had been active in the BLA and were genuinely dedicated to its continuation; indeed, some felt more accountability to miners and black lung activists in the coalfields than to their employers in Washington. During 1973, Fritts and others at the Field Service Office helped to organize regionwide black lung meetings at which a common organizational agenda was hammered out and a regional BLA structure built. The staff also made the office a conduit for communicating BLA needs and demands to appropriate individuals in the union and worked to ensure BLA representation at key UMWA events, such as the Pittsburgh convention. Toward the end of 1973, Miller added to the staff a lawyer, Gail Falk, whose work on the legal aspects of black lung compensation made the office an indispensable source of information, especially for the lay advocates.

As the credibility of the Miller administration declined and the fragmentation of the Black Lung Association increased, however, the position of the Field Service Office became more and more untenable. It was shelled from all sides, caught in a political no man's land between warring factions in the United Mine Workers and the Black Lung Association. Amid the discord, it became impossible to determine who, if anyone, represented the legitimate voice of the BLA. When the Field Service Office staff worked with one faction, they were attacked by the others. When they threw the institutional weight of the union behind the BLA's legislative goals, they were accused of usurping leadership and control over the black lung issue. Critics within the UMWA also assailed the Field Service Office. As wildcat strikes rocked the coalfields of southern West Virginia, Miller became increasingly paranoid about his loss of control over the union's internal machinery and its rank and file. He concluded that the Field Service Office was an instigator of rebellion and, in the summer of 1975, ordered it closed.

Apart from the intrigue and factionalism that characterized relations between the UMWA and BLA, the black lung movement was wracked by other internal divisions. Basic questions concerning goals and strategies, on which activists again split into opposing camps, arose in the years after 1972. Many of the longtime leaders and lay advocates adhered to the traditional concept of the Black Lung Association: an organization dedicated to claims assistance and legislative reform of the federal compensation program. Although they and other activists concurred about the importance of disease prevention and the strategic need to reach working miners, no one implemented this recognition through aggressive organizing. These individuals tended to lobby cooperatively with the union for black lung reform, but they often found themselves at the rear, rather than at the forefront, of the legislative campaigns. The UMWA's extensive institutional assets—finances, lobbyists, technical expertise, and ability to mobilize working miners—completely overshadowed the meager resources of the declining Black Lung Association. One union staffer commented:

> A lot of the black lung people thought that the UMW was taking over. But if your goal was a black lung reform law, then going through the union and the working miners was the only way you had a chance. That was a lot of my debate with the Black Lung Associations those years—getting working miners out. Which is what the BLA people always said they wanted. But when you did that, the working miners were always so much more strong and vital, it kind of left black lung people in the dust.[33]

Black lung actvisits with a different political vision stood more aloof from the UMWA-orchestrated lobbying. For these individuals, improvement of the compensation program was only one of many political goals, and legislative lobbying was a tactical maneuver in a larger struggle. They tended to see the private ownership and dominant power of the coal industry as the ultimate source of exploitation and poverty in the coalfields. They condemned the Miller administration for betraying the hopes of MFD and for attempting to suppress a rank-and-file movement that could take on the industry at the bargaining table and in the workplace, ultimately in the form of wildcat strikes. They sought to link the Black Lung Association more directly to working miners and to build a unified power base that could enforce the accountability of both union leaders and industry officials. Those with this political view included most of the newer, young organizers, as well as a few indigenous radicals whose experiences as workers and residents of the coalfields had yielded a potent blend of class consciousness, anti-authoritarianism, and die-hard independence.

Although some activists had participated in the successive campaigns to reform the black lung compensation program, they were growing impatient with this singular focus. One disabled miner argued:

> You hear so much talk about black lung, black lung. If they want to do something about black lung, they'll have to start in underground where the disease really starts. They don't want to compensate people for their health; they don't want to pay you if you get the disease. But really, if they *did* pay everybody for black lung that are entitled to it, really all they're doing is petting the disease. They're just nursing it. To do something, they've got to start where the black lung starts—in the mines.[34]

Hoping to broaden the scope of the movement to include issues like prevention, these activists resisted the tendency to concentrate all their political energy on lobbying in Washington for improved compensation. Describing this political difference within the movement, one miner commented:

> What it was was a difference in philosophy: was the Black Lung Association going to be a lukewarm, reasonable organization that goes for lobbying, or was it going to be a militant, fighting organization? There was a lot of struggle around lobbying. . . .
>
> The active miner, the working miner, is the key; they are pivotal in any struggle in the coalfields. The disabled and retired people don't have a lot of power except lobbying. They lobbied the wrong

people, though. They lobbied Congress instead of the rank-and-file miners.[35]

These conflicting perspectives on the goals and strategy of the black lung movement collided during the campaign for automatic entitlement to federal compensation. In 1973, those attending a regional BLA meeting in Pikeville, Kentucky, formally adopted automatic entitlement to black lung compensation after twenty years in the mines as a legislative goal.[36] Frustration with the aftermath of the 1972 amendments, especially the high denial rate under the Labor Department's black lung program, generated this provocative demand. Eligibility based on medical determinations had not yielded a fair and consistent (much less "scientific") program, but instead had left compensation up to the influence and predisposition of political appointees and administrators. Because there was no clear-cut diagnostic process on which all could agree, automatic entitlement was the most simple and equitable alternative. It would align eligibility criteria with the activists' conviction that black lung afflicted most people who had worked a substantial period of time in the mines.

BLA unity around the automatic entitlement demand collapsed after activists began to lobby and politic on Capitol Hill. Disputes arose over the role of the UMWA, which orchestrated an effort by lobbyists and a mass rally of coal miners in support of automatic entitlement. Disagreement also arose over the demand itself: from the start, some assumed that automatic entitlement was impossible to achieve but promoted it as a lever to pry from Congress more liberal eligibility standards; others took it seriously as an actual goal. When, in the early months of 1976, the U.S. House Committee on Education and Labor reported out a bill, H.R. 10760, granting limited automatic entitlement, activists split over whether to accept the compromise. A few members of the Beckley, West Virginia, Black Lung

Association denounced the bill as "giving us nothing" and called on miners to walk out if the demand for fifteen years "equals benefits" and additional demands were not met.[37]

Those who ridiculed this bill and agitated for rank-and-file action against it soon found themselves opposed and isolated by other black lung activists. Some saw the legislation as the culmination of three years of effort in Washington and pinned their hopes on its passage; those who were more skeptical of the legislative reform still believed that the Beckley group had jumped the gun in calling for strike action. They argued that the timing was poor, given the stage of the legislation, and they believed that a strike was premature in terms of their educational and outreach effort among working miners. Those who advocated a strike tended to view this opposition as "backward" and proceeded with their rallies, leafletting, and other agitation. Amid a barrage of telegrams, radio announcements, and other public statements by UMWA officials urging miners not to stop production, on March 1, 1976, thousands in southern West Virginia walked out. Within a few days, however, the strike began to fizzle, and by the following week, all miners were back on the job.[38]

Timed to coincide with a vote on H.R. 10760 by the House of Representatives, the wildcat apparently had little impact in this arena, for the legislation passed on March 2, 1976. For working miners, the strike may have confused more than it educated, because the conflicting statements of black lung activists left many miners unsure whether they were being asked to support or oppose the legislation. The wildcat further discredited UMWA president Arnold Miller, whose no-strike orders were easily ignored. But in retrospect, the strike resulted above all in the defeat of efforts to revitalize and redirect the black lung movement by irreparably dividing activists who hoped to accomplish this. The unmistakable decline of the movement made success unlikely, but the black lung activists' inability to unite made it utterly impossible.

## The End of a Movement

Social movements do not begin and end according to identifiable schedules; the edges of their history are often ragged and ill defined. Selecting a single point in the black lung movement's long process of disintegration and declaring it to be the "end" is ultimately a rather arbitrary exercise. By 1973, some already believed that the movement had run its course, but it was not until four years later that other participants and observers acknowledged that it was over. After the battle for automatic entitlement was lost in the U.S. Senate during the last months of 1976, many of the remaining activists admitted defeat. By 1977, the black lung movement was no longer a political force in the coalfields.

Remarkably, legislative reform of the federal black lung program was not over. In 1978, President Carter signed into law a new Black Lung Benefits Reform Act that required more liberal medical eligibility criteria, prohibited the rereading of X-rays under most circumstances, and eased the claims process for working miners and miners' widows.* Companion legislation established a trust fund, financed by a special tax on coal, to pay the claims of certain individuals. With enactment of the reforms, the claims approval rate soared, and by the end of 1979, over forty thousand additional miners and widows had been awarded compensation.[39]

Political considerations similar to those that obtained in 1972 were partly responsible for the legislation's passage. But the influence of the black lung movement, which extended beyond its own life span, was also at work. The movement continued in

* These reforms have now been reversed. See chapter 7.

the form of UMWA staff members who took up the banner of black lung reform and lobbied for the new legislation. The movement's influence also persisted through a few stalwart lay advocates who made the rounds on Capitol Hill and maintained the image of an organized bloc of pressure. Most importantly, the movement lived on in the rank-and-file upheaval that the wildcat strike over black lung disease had initiated nine years earlier. During the winter of 1977–78, while the U.S. Congress debated black lung reform, coal miners were deadlocked with the operators in the longest contract strike in UMWA history. This would be the final showdown of the reform period.

## The End of an Era

By 1977, massive, uncontrollable wildcat strikes had shattered Miller's credibility with the operators and united them with an intractable determination to impose "stability" on their labor force. In a pronounced shift from previous strategies, the industry went on the offensive with the 1977 contract negotiations, hoping to boost declining productivity with incentive schemes, to arrogate to itself the power to impose labor discipline by firing wildcat strikers, and even to abolish one of the union's greatest sources of pride and appeal, the Welfare and Retirement Fund.

During collective bargaining in 1974, the first for the union's reform administration, many member companies in the Bituminous Coal Operators Association clearly had hoped to buy labor peace with a generous settlement. Not only did the 1974 contract significantly advance miners' economic interests through a cost-of-living allowance, sick pay, a big pension boost for recent retirees, and other provisions, it also affirmed the MFD commitment to improved occupational health and safety by requiring helpers on certain equipment, granting individual miners the right to withdraw from imminently dangerous work situations, and expanding the powers of the local safety committee. Industry negotiators believed that they had gone more than halfway in meeting miners' demands; as a quid pro quo, they expected the turmoil in their work force to subside. As *Business Week* observed in 1974:

> From the operators' point of view, there was hope that the goodwill created in the negotiations would result in union-industry cooperation in reducing absenteeism, halting wildcat strikes—although this tradition would die slowly—and increasing productivity. The BCOA paid a healthy price, at a gamble, to achieve this, but there was no doubt that beginnings had been made.[40]

The BCOA lost its gamble and lost it badly. Industrial relations only worsened in the years that followed the 1974 contract. Some rank-and-file miners looked at the industry's bountiful profit picture and the paltry contractual gains of previous years and concluded that the 1974 settlement was not enough. Resentment also flared over the accelerated ratification process, which the contract squeaked through by a margin of only 57 percent. In the first months under the new agreement, many miners became convinced that the operators were trying to take back in the workplace and through the grievance procedure what they had won at the bargaining table. With record-breaking frequency, miners protested by refusing to enter the mines. In 1975 and 1986, there were more than twenty-four hundred wildcat strikes in bituminous coal.[41] Instead of labor peace, the operators had gotten "industrial anarchy."[42]

In the farsighted view of history, these wildcats represent the climax to the rank-and-file movement that began in 1968. Miners had run through the course of established mechanisms for reform: they had campaigned twice for power in the union, lobbied for state and federal legislation concerning occupational safety and health, pursued their grievances through contractual procedures and the courts, and negotiated their demands at the bargaining table. Yet despite legal, legislative, and institutional victories, their goals of a union genuinely run by the rank and file and of a workplace relatively free of danger remained beyond their reach. In 1975–76, miners abandoned legal and institutional mechanisms and faced off directly with the operators in an unmediated confrontation. Industry efforts to force miners back to work through court injunctions only fueled the wildcats and transformed them into political struggles against the judicial system and its role in enforcing labor discipline. In the summer of 1976, a four-week wildcat that eventually brought out one hundred thousand miners finally forced the operators to retreat from their use of the courts.[43] But it also galvanized their determination to find other punitive means to subdue their work force.

The operators came to the bargaining table in 1977 with a sheaf of take-back demands, fully prepared to outlast miners in a lengthy strike, having little intention to compromise. In the 1974 contract, they felt they had offered the carrot; now they would use the stick. Lacking any pretense of leadership at the top of their union and in many cases in serious financial need, miners nevertheless held out against this offensive in a nationwide strike that lasted almost four months. Twice UMWA negotiators were sent back to the bargaining table, their proposals scorned by the union bargaining council and then the rank and file. But industry intransigence and the ineffectuality of the union leaders gradually wore away miners' hope that further sacrifice would yield a substantially improved settlement; on March 24, 1978, miners ratified by a bare margin the third proposed contract. Although it eliminated some of the most humiliating provisions of the two tentative documents that had preceded it, the 1978 contract was profoundly, unmistakably regressive. Most important, this document demolished in a few pages the medical care program of the Welfare and Retirement Fund, replacing it with private insurance coverage. Through an appended memorandum of understanding, the contract also gave companies the right to fire workers who participated in or even verbally supported wildcat strikes.

Subsequent developments in the coalfields suggest that the 1978 contract represented more than a temporary setback or even forced retreat for miners: it marked the end of the era of reform. In 1979, Arnold Miller formally turned over the union presidency to his vice-president, Sam Church, a former Boyle supporter, in a move that merely ratified internal power changes that had already occurred. The successful industry offensive has persisted. Today, punitive actions against miners who protest deteriorating working conditions or other problems combine with heavy layoffs in the deep mines of Appalachia to give an uneasy reality to the rhetoric of "labor peace." Encouraged by the rank-and-file weakness, some operators are going in for the kill: nonunion companies are penetrating even the UMWA's stronghold in southern West Virginia.

Achievements within the union are evident, however, and they bear witness to the continuing, positive impacts of the reform era. A new generation of youthful leadership, exemplified by current UMWA president Rich Trumka, has risen to power. After a period of retrenchment under Sam Church, there has been a renewal of the union's commitment to occupational health and safety and certain other issues raised by the reform movements. Democratic provisions, formalized in the constitution, guarantee rank-and-file members a vote in the election of their leaders and a voice in the union's affairs. Those who spent years fighting for the cause of union democracy can take heart from these institutional reforms. Yet, sadness lingers among those who remember the vision to which they once aspired:

> I think a lot was accomplished. But I don't think that's the question. The question is what the *possibility* was. There was a possibility in my mind of turning around the whole labor movement, of mak-

ing a start at that. The Miners for Democracy was a genuine rank-and-file movement. Arnold Miller, when he was elected, was a genuine rank-and-file leader. He had a very good staff that was willing to work day and night, totally committed to him and to those ideals, which is a rarity in this town. There were lots of good ideas, a constituency that was ready to be educated, ready for change, ready to support meaningful and militant action on the part of the international union. That got started, and it died stillborn. So *that's* what the tragedy is. It was the possibility. I don't know a person who was involved who doesn't still live with that every day.[44]

7

# Whose Body?

The black lung movement broke through the wall of silence that concealed the occupational illness and disease of workers in the United States. It initiated a controversy over occupational health that continues in many industries to this day. Textile workers fashioned a brown lung movement after coal miners' example. Asbestos workers pressed for a federal compensation program similar to that covering disability and death from black lung. With increasing frequency, trade union activists included occupational safety and health protection among their collective bargaining demands; in the legislative arena, they secured a federal Occupational Safety and Health Act that was patterned in part after the Coal Mine Health and Safety Act of 1969.

For its constituents, the black lung movement achieved a unique and unprecedented federal compensation program. Half a million miners and widows have received compensation under the federal black lung program; especially for those ineligible for a pension or other benefits, the monthly payments have meant the difference between destitution and a modest survival.[1] The respirable dust standard and other disease prevention measures incorporated in the U.S. Coal Mine Health and Safety Act of 1969 were also attributable to the black lung movement. Finally, as one element in a larger upheaval throughout the coalfields, the black lung movement contributed to the rank-and-file takeover of the United Mine Workers of America and to the democratization of the union's internal structure.

Originally and essentially, however, the black lung movement was a struggle over the recognition of an occupational disease. What seemed at first to be a straightforward task—achieving legal inclusion of a "new" dust disease under the workers' compensation system—turned out to be a far more complex undertaking. "Black lung" as refracted through the lens of scientific medicine was quite different from the disease for which miners sought recognition and compensation. In a struggle that lasted nearly ten years, activists persistently challenged physicians and policymakers over the meaning of this disease; for a time, they were able to replace the restrictive scientific construction of coal workers' pneumoconiosis with their own definition of "black lung." Although focused on seemingly technical disputes over diagnostic methods, disability standards, and other issues, this conflict over the definition of black lung was intensely political: it involved the conceptual limitations of scientific medicine, the ideological content of medicine's view of disease, and the powerful role of physicians in labeling work-related disability as "legitimate." It raised the unsettling possibility that the knowledge base of scientific medicine is neither conclusively true nor politically neutral.

## The Antiseptic Physician

According to the dominant viewpoint of our culture, scientific knowledge is untainted by social influences or the personal beliefs of its creators.[2] It is regarded as "the only truly valid type of knowledge."[3] Many who study the history and philosophy of science argue that science enjoys this special epistemological status not only because of its method (which may be applied in the "soft" social sciences without the same claim to factualism), but also because of its object: the natural world. The "laws" of nature are uniform across space and time; although change within a limited range is always taking place, nature is permanent. Because of these qualities, which stand in contrast to the diversity and discontinuity of human history, nature is knowable in an ultimate sense.

Counterpoised to adherents of this traditional perspective is a growing chorus of critics who argue that science is not an autonomous realm insulated from social life by the purity of its internal logic or the permanence of its external object.[4] From the Copernican revolution to the development of space shuttles, from academic training to the organization of commercial laboratories, the practice and knowledge of science have been and are shaped by the social context.

Sociologists, for example, have analyzed the mundane transactions of the laboratory, where scientists may be observed negotiating over the conduct of experiments, the acceptability of certain evidence, and the interpretation of findings in ways that guarantee the social contingency of "purely technical" findings.[5] Feminists have examined not only the male-dominated organization of science, but also the gender-laden character of certain scientific concepts. The concept of objectivity, for example, with its masculine connotations of autonomy, distance, the radical separation of subject and object, has been scrutinized in both historical and psychoanalytical terms.[6] In critical reviews of contemporary scientific research, practicing scientists, primarily biologists, have documented the deeply entrenched tendency to interpret physical processes as individualistic and hierarchical, and in other terms not "required" by the available data but commensurate with existing social arrangements.[7]

Scholars have also investigated the historical relationship between the birth of modern science in the seventeenth century and the growth of capitalism in western Europe.[8] This relationship involved far more than the overthrow of the secular power of the Church, and the resultant freedom to espouse nonreligious, scientific concepts; the emergent social relations of capitalism made possible new forms of perception and conceptualization, which were critical to the development of scientific thought. The transformation from a rigid and static social order seemingly ordained by God to a mobile society of free, acquisitive individuals, shorn of obligatory ties to land or feudal rank, was implicated in the developing scientific approach to nature: no longer a process of passive reflection, the new method was to intervene, manipulate, and, ultimately, dominate the natural world. The growth of economic production that depended on the "freedom" and interchangeability of human labor power (and, in the philosophical realm, affirmation of the essential equality of all human beings) was reflected in and reinforced by the reductionist scientific effort to ascertain the common essence of matter: absolute masses, atoms, cells. New perceptions of change, mutability, and time (now no longer circular, but progressive) may also be seen as social derivatives, yet they were essential to scientific thought—for example, in the biological theory of evolution.[9]

The political implications of this emerging body of literature, which from different perspectives explores the social character of science, are considerable. Insofar as scientific knowledge incorporates elements of social beliefs, however implicit, obscured, or abstract, it follows that science tends to legitimate those beliefs by investing them in the natural world. At the same time, however, the ideology that surrounds the scientific enterprise—that of an autonomous and exclusive quest for absolute truth—admits no external criticism, much less alternative, nonscientific explanations of "natural" phenomena. The result is not only a hermetic, self-legitimating system of scientific belief, but also virtually unassailable ideological authority. Insofar as science alone appears to reveal the immutable facts and laws of nature, its hegemony is even more complete than the religious world-view that it replaced. To challenge God was merely foolish; to challenge nature is impossible.[10]

Viewed within a framework that affirms the social character of science, the black lung controversy acquires deeper meaning. It was not just a conflict stimulated by medical uncertainty and disagreement, with conservative physicians on one side and militant coal miners and their medical allies on the other. The perceptions of black lung that those on opposing sides brought to this controversy were shaped by their distinctive class positions and social experience. The result was an ideological struggle between two different views of the world, both of which were intensely political. At issue in that struggle was not only the definition of a single occupational disease, but ultimately the scientific construction of all disease, and the social role of physicians in controlling and applying that definition.

The prevailing medical view of disease is of a biological event involving an aberration or aberrations in the structures and functions of the body. According to Steadman's *Medical Dictionary*, disease is "an interruption, cessation, or disorder of body functions, systems, or organs."[11] Implicit in this functional view is the conception of the body as a system of mechanically related parts—a machine. This approach, in which the mind is separated from the body, arose as part of an overarching mechanistic paradigm within the natural sciences; it both made possible and was reinforced by the development of the machine as a major force of production and the development of "free" human labor as an appendage to that machine.[12] The conceptual power and practical applications of the mechanistic understanding of the body are immense. A recent and vivid example is the surgical implantation of the artificial heart, in which a body is disassembled, fixed with a mechanical replacement part, then put back together again. At the same time, however, the mechanistic paradigm is implicated in many of the major limitations of scientific medicine: the treatment of the patient as an object—indeed, the treatment of disease rather than patients; the functional definition of a healthy human body as a working machine; the failure to conceptualize dynamic interactions within the whole person (mind and body) in the process of disease; and the emphasis on curative techniques, including surgical intervention, rather than on disease prevention.[13]

Scientific medicine situates disease spatially, within the individual body, and temporally, at the point when signs, symptoms, or other physical alterations develop. Clinical medicine reflects this understanding of disease on a practical level: individual patients present the physician with their distinctive symptoms and complaints; they appear as random, disconnected "cases," and they are granted therapeutic treatment as individuals. There is no social meaning to disease in the sense of an internal relationship between social structure and the individual experience of ill health. Disease is ahistorical as well as asocial; it has no history except a "natural," that is, physical, history. It is said to exist when experienced by the individual and diagnosed by the physician, not at the point when it is being produced (as in the coal industry during the 1950s). The possibilities for prevention are thus constrained within the very definition of disease.[14]

At the level of causation, germ theory, with its focus on discrete, physical agents, complements and reinforces the scientific construction of disease as an individual, biological event. When it first arose during the last decades of the nineteenth century, germ theory appeared to obviate the need for "unscientific" (i.e., social, nonphysical) explanations of disease causation, and it was used as an ideological weapon to defeat the proponents of social medicine.[15] More recently, germ theory has come under attack for its failure to elucidate the precise origins of the most significant diseases of our place and time—heart disease and cancer. Medical textbooks now routinely allude to the "multifactoral" nature of disease causation. This merely states the problem, however; it does not solve it. The multiple causal factors of a chronic disease are typically an ad hoc list of individual and environmental attributes—cigarette smoking, obesity, air pollution—which exhibit no internal coherence or relationship to one another. Multifactoral causation thus provides no conceptual framework for investigating the simultaneous interaction of social and / or biological factors. Germ theory remains the operational paradigm of the research laboratory.

In the course of the black lung movement, this scientific view of physicians, who regarded disease as an individual, biological event associated with a discrete physical agent, collided with miners' perceptions of their own experience. Their bitter controversy devolved on several levels at once. Miners and other activists contested the narrow, scientific construction of black lung by insisting on the legitimacy of their own experience. In so doing, they implicitly challenged the exclusive ideological authority of medicine to control the definitions of disease and disability. Further, activists contested the assumed neutrality of physicians who utilized their professional authority to support a restrictive interpretation of coal workers' pneumoconiosis; this apparently neutral scientific stance was, in both content and implication, intensely political. Finally, the black lung movement challenged the economic role of physicians in regulating the labor force through their power to pronounce some workers legitimately "disabled" and others (at least in a functional sense) "healthy." These challenges to physicians did not arise from fundamentalist religious beliefs or other anti-scientific sensibilities, but from the anger of individuals who believed that the superior legitimacy automatically granted scientific medical knowledge represented a complex and powerful form of social control.

From the perspective of coal miners, the restrictive medical view of black lung reconstituted their collective experience of exploitation and disease as an individual, clinical diagnosis that physicians controlled. Miners' conviction that they were disabled by a pervasive black lung disease became, in the eyes of physicians, the subjective impression of patients; black lung was in fact coal workers' pneumoconiosis, a single clinical entity, disabling only in relatively advanced (and rare) stages. The disease acquired legitimacy—indeed, came into existence—only when visible to trained personnel viewing objective diagnostic evidence, that is, X-rays, of individual miner's lungs. The thousands of older miners who believed themselves disabled by black lung yet exhibited no X-ray evidence of advanced CWP might legitimately be considered "disabled"—if the quantitative results of certain tests confirmed such a condition. However, the origin of their disability was nonoccupational, above all their own cigarette smoking. Although this scientific definition of disease was quite different from physicians' earlier construction of "miners' asthma," the result, in the eyes of the victims, was the same: black lung was trivialized. What miners viewed as a collective problem became, from the perspective of scientific medicine, individual, quantifiable cases. What they experienced as part of the shared social world of coal mining became occasional, biological events. What they attributed to their class relationship with the coal operators became the product of a single physical agent, dust. In sum, what was collective became individual, what was social became biological, what was meaningful became accidental, what was a produced condition became a natural, physical phenomenon.

Embittering the medical controversy over black lung was the professional authority of physicians to define experiences of which miners had direct and intimate knowledge—and physicians had none. In performing physical examinations for black lung compensation, physicians' professional activities included diagnosing miners' diseases, identifying their origins (especially occupational versus nonoccupational), and quantitatively assessing any associated disability. In the miners' view, physicians possessed the authority to declare whether or not they were sick, what made them sick, and how sick they were. In quantifying disability and allocating it to occupational or nonoccupational sources, physicians implicitly assessed the conditions in which miners had lived and worked all of their lives. That most physicians had never been in a coal mine (much less worked in one), and that some had never even been in the coalfields, served to intensify the conflict between physicians and those whose experiences they had the power to quantify and define.

The authority of physicians to pronounce miners as "healthy" or "disabled" endowed them with power that was not only cultural and psychological; it also carried important financial consequences. In the context of federal black lung compensation, doctors acted as gatekeepers: they controlled access to financial benefits that were allocated according to medical eligibility criteria. Although administrative and, ultimately, judicial personnel actually made the decision to award or deny benefits, medical evidence was often a determining factor in the outcome. Many claimants logically though incorrectly believed that physicians directly controlled this decision. The perceived (and, to a lesser extent, actual) power of physicians over claimants' financial future further complicated and embittered the black lung controversy.

The tug-of-war over financial benefits also had a deeper meaning in that it involved conflict over the structural economic role of physicians. Doctors act as gatekeepers in a more generic sense than guarding eligibility for financial benefits: they control access to the "sick role", the sole avenue by which adults may legitimately escape the daily responsibilities of their class and gender.[16] For coal miners, as for other workers, the preeminent requirement of their class position is to perform wage labor. Medical criteria for assessing disability (and determining compensation eligibility) that take as the standard for "health" the functional capacity to work explicitly enforce this requirement. Even if damaged by their work, coal miners still must provide medically sanctioned evidence of their "total disability"—i.e., complete inability to continue functioning as coal miners—in order to receive financial compensation and relief from wage labor. In pushing against the limits of this compensation policy, miners contested not only the ideological authority of physicians in defining disease and assessing disability; they ultimately threatened the economic power of the coal operators by pressing for a broad definition of black lung and a relaxed standard of disability that would provide unhealthy miners an alternative to labor in the mines.

The coincidence between the restrictive scientific view of black lung and the economic interests of the coal industry was, for miners, an ultimate source of distrust and conflict with physicians. The narrow definition of disabling black lung as a relatively rare, complicated pneumoconiosis was highly functional to the industry: it circumscribed the scope of occupational lung disease and correspondingly diminished both the cost of compensation benefits and the importance of prevention. In a more subtle fashion, it reconstituted an explosive political issue as an individual, biological event. In the context of policy formation, the scientific knowledge of medicine played a mediating role between the interests of the coal industry and the actions of the state. It facilitated distance between these economic and political institutions, and appeared to ground policy in the neutral, technical knowledge of a third party.[17]

Physicians who advocated a strict definition of compensable disease were not simply "company doctors," however, nor were they necessarily political conservatives.[18] Their understanding of black lung was invested with political content not because of their individual affiliations and loyalties, but because the scientific knowledge that informed their understanding was itself a social product. The development of scientific knowledge and the articulation of capitalist social relations have been internally related historical processes. They have generated and legitimated mutually reinforcing views of the world—a world broken apart into discrete, individual components, whose essential character can be studied, quantified and discerned apart from that of all others. The meaning of life is sought in the molecule; the meaning of human society is located in the individual.[19] That the relationship between the knowledge of science and the domination of capitalism is oblique, complex, and largely invisible renders it the more powerful. The final, continuing challenge of the black lung movement thus assumes the form of a question: what kind of body does capital require—and medicine enforce—among coal miners and all other workers in our society?[20]

## Epilogue

More than a century and a half have passed since Dr. James Gregory opened up the lifeless body of John Hogg and hypothesized a connection between the miner's blackened lungs, his respiratory disability in later life, and his occupation. For a time, physicians in Britain and the United States continued to investigate the relationship between occupational exposures and miners' respiratory disease. Toward the end of the nineteenth century, however, during a period of tight corporate control in the Appalachian coalfields and an increasingly restrictive scientific definition of disease, black lung disappeared from the medical literature of both countries. In the U.S., coal miners eventually precipitated renewed medical attention to black lung by winning a union-controlled health care plan for themselves and their families. Even so, coal workers' pneumoconiosis—much less black lung—was not accepted as a legitimate occupationally-related disease by the medical profession as a whole. Formal recognition again required the active intervention of coal miners themselves.

Even as social factors have impinged on the medical construction of black lung, so have they shaped the actual production of disease. Black lung originated not simply from the physicial presence of dust in coal mines, but from the relative power and respective actions of miners and operators, which influenced conditions in the workplace. During the craft era of coal mining, social factors relevant to the production of disease included cost-conscious operators' failure to ventilate mines adequately, miners' unsteady grip on unionization, and the frequent lowering of the piece rate, among others. Miners' eventual success in unionization enhanced their collective power in the workplace, but in paradoxical ways it also undermined their capacity to make that workplace healthy and safe. In the years after World War II, fundamental changes in the structure of the industry and the role of the United Mine Workers produced unimpeded mechanization of the production process, high levels of unemployment, forced migration, and occupational death and disability from black lung. Ultimately, these changes also produced a situation in which overt cooperation between the union and the industry was neither tolerable for the rank and file nor useful to the operators. There ensued a decade of struggle over their new relationship.

This struggle involved numerous issues, different arenas, diverse institutions, and thousands of individuals. At the center of the controversy, however, were the unifying questions of who and what terms would define class relations in the coalfields. For activists in the black lung movement, these questions assumed the form of who and what terms would define black lung. For a time, rank-and-file miners and their families took the offensive. They insisted upon recognition and remedial action for the occupational dangers of coal mining, reclaimed control of the United Mine Workers, and pressed the operators for collective bargaining concessions. Dissatisfied with the limitations of these accomplishments, they persistently sought to secure and defend their goals through additional legislative reforms and, ultimately, through a blaze of wildcat strikes.

Today, the situation in the coalfields is altogether different. Economic despair has settled in the region. Layoffs in underground mining grant West Virginia the dubious distinction of one of the highest unemployment rates of any state in the nation. Technological innovation in the form of longwall mining machinery, determined productivity drive by the operators, and a depressed market for certain grades of coal combine to produce the layoffs. Fear of job loss, a large pool of surplus labor, and the aggressive anti-union stance of companies like A. T. Massey undercut the leverage of the rank and file. Once again, the workplace is being reorganized and new technology introduced at a time of rank-and-file weakness. Even more than continuous miners, the highly productive longwall machines stir up thick clouds of dust. Black lung disease awaits the younger generation of coal miners who are now at work underground.[21]

At the same time, the official definition of black lung that guides the federal compensation program has once again been narrowly circumscribed; the problem of occupational lung disease, and financial liability for it, has been reduced accordingly. In 1980, the U.S. General Accounting Office reviewed a random sample of approved black lung claims, and concluded that "in 88.5 percent of the cases, medical evidence was not adequate to establish disability or death from black lung."[22] In the same year, physicians at the West Virginia University Medical Center reported the results of their clinical study of respiratory disability among 150 coal miners: none had progressive massive fibrosis (associated with the advanced stages of CWP), which they asserted was "the only true disabling form of coal workers' pneumoconiosis."[23] In related comments, the editor of the *Jour-*

*nal of the American Medical Association* suggested that the prospects of receiving financial benefits "create symptoms" among coal miners. The U.S. General Accounting Office subsequently recommended that the U.S. Congress "redefine black lung as coal workers' pneumoconiosis", eliminate the legal presumptions that assisted claimants in establishing eligibility, allow the U.S. Department of Labor to re-read X-rays, and make other changes that would restrict awards to those "legitimately" disabled by CWP.[24] In December, 1981, the Congress passed legislation that adopted several of the GAO recommendations. The great majority of new black lung claims (as of the last official report, 94%) cent) are now denied.[25]

This state of affairs does not diminish the significance of what miners and their families were able to accomplish in the black lung movement. Indeed, the present reaction bears witness to the power of those economic and ideological forces that, for a time, they were able to override. Their victory yielded not only recognition and compensation for occupational lung disease, but also the revelation of a complex relationship between the ideological content of medicine and the economic power of capital. That the underlying conflict between miners and operators will break out again in future movements is certain. Ever since the first investors laid claim to the coal of Appalachia, the people of this region have been revolting in various forms against the appropriation of their land and their labor by an industry that they do not control. Those who fought in the black lung controversy are both heirs of and contributors to this long history of resistance. Their movement joins a tradition of efforts by miners and their families to reclaim their land, their work, and their health as genuinely their own.

# Notes

## Chapter 1

1. James Craufurd Gregory, "Case of peculiar Black Infiltration of the whole Lungs, resembling Melanosis," *Edinburgh Medical and Surgical Journal* 36 (1831): 391.

2. Ibid., p. 392.

3. William Marshall, "Cases of Spurious Melanosis of the Lungs, or of Phthisis Melanotica," *Lancet*, 1834, pp. 270–74; Matthew Gibson, "On the 'Phthisis Melanotica' (so called) of Coal-Miners," *Lancet*, 1834, pp. 838–39; William Marshall, "Remarks on Spurious Melanosis of the Lungs," *Lancet*, 1834, pp. 926–28. For historical summaries, see George Rosen, *The History of Miners' Diseases* (New York: Schumans, 1943); and Andrew Meiklejohn, "History of Lung Diseases of Coal Miners in Great Britain: Part 1, 1800–1875," *British Journal of Industrial Medicine* 8 (July 1951), "Part 2, 1875–1920," *British Journal of Industrial Medicine* 9 (April 1952); and "Part 3, 1920–1952," *British Journal of Industrial Medicine* 9 (July 1952).

4. "Report of the Schuylkill County Medical Society," *Transactions of the Medical Society of Pennsylvania*, 5th ser., pt. 2, June 1869, p. 489.

5. Ibid., p. 491.

6. See the articles cited in note 3 from the 1834 edition of *Lancet*. For the United States, see Andrew Roy, *History of Coal Miners of the United States* (Columbus, Ohio: J. L. Trauger, 1907), pp. 118–20, 444–52; and Henry C. Sheafer, "Hygiene of Coal-Mines," in *A Treatise on Hygiene and Public Health*, ed. Albert H. Buck (New York: William Wood, 1879), 2:229–50.

7. Friedrich Engels, *The Condition of the Working Class in England*

(Oxford: Alden Press, 1971), pp. 280–81.

8. Émile Zola, *Germinal* (New York: E. P. Dutton, 1933).

9. Dr. William Anderson, as quoted in the Charleston (W. Va.) *Gazette*, April 16, 1969.

10. Andrew Meiklejohn, in his three-part series on the "History of Lung Diseases of Coal Miners in Great Britain," argues that technological innovations during the late nineteenth and early twentieth centuries improved underground working conditions and reduced the incidence of occupational disease; in his view, this explains the medical inattention. The situation in Great Britain is beyond the scope of the present work. However, for the United States, there is no evidence to suggest that working conditions improved during this period. If anything, they may have worsened due to the pressure surrounding productivity, the intense competition within the industry, and an influx of untrained workers. See chapter 2.

11. Michel Foucault, *The Birth of the Clinic: An Archaeology of Medical Perception* (New York: Vintage Books, 1975); and N. D. Jewson, "The Disappearance of the Sick-Man from Medical Cosmology, 1770–1870," *Sociology* 10, no. 2 (May 1976): 225–44.

12. Erwin H. Ackerknecht, *Medicine at the Paris Hospital, 1794–1848* (Baltimore: Johns Hopkins University Press, 1967); Ivan Waddington, "The Role of the Hospital in the Development of Modern Medicine: A Sociological Analysis," *Sociology* 7 (1973): 211–24; Charles E. Rosenberg, "And Heal the Sick: The Hospital and the Patient in the 19th Century America," *Journal of Social History* 10, no. 4 (1977): 428–47

13. X. Bichat, *Anatomie Generale*, as quoted in Foucault, *Birth of the Clinic*, p. 146.

14. Rosenberg, "And Heal the Sick."

15. "Report of the Schuylkill County Medical Society," p. 490.

16. Dr. D. C. Rathburn, as quoted in Sheafer, "Hygiene," p. 246.

17. Jewson, "Disappearance of the Sick-Man"; Meredeth Turshen, "The Political Ecology of Disease," *Review of Radical Political Economics* 9, no. 1 (1977): 45–60; Karl Figlio, "The Historiography of Scientific Medicine: An Invitation to the Human Sciences," *Comparative Studies in Society and History* 19 (1977): 262–86; Howard S. Berliner and J. Warren Salmon, "The Holistic Health Movement and Scientific Medicine: The

Naked and the Dead," *Socialist Review* 9, no. 1 (January–February 1979): 31–52.

18. Jewson, "Disappearance of the Sick-Man."

19. Paul Starr, *The Social Transformation of American Medicine* (New York: Basic Books, 1982).

20. Ronald D. Eller, *Miners, Millhands, and Mountaineers: Industrialization of the Appalachian South, 1880–1930* (Knoxville: University of Tennessee Press, 1982); John Gaventa, *Power and Powerlessness: Quiescence and Rebellion in an Appalachian Valley* (Chicago: University of Illinois Press, 1980); Harry Caudill, *Night Comes to the Cumberlands: A Biography of a Depressed Area* (Boston: Little, Brown, 1962).

21. Eller, *Miners, Millhands, and Mountaineers,* p. 64.

22. Ibid.; and David Alan Corbin, *Life, Work, and Rebellion in the Coal Fields: The Southern West Virginia Miners, 1880–1922* (Chicago: University of Illinois Press, 1981).

23. U.S. Coal Commission, *Report* (Washington, D.C.: U.S. Government Printing Office, 1925), p. 1431.

24. Senate Committee on Education and Labor, Subcommittee, Acting under Senate Resolution 37, *Conditions in the Paint Creek District, West Virginia,* Hearings, 63d Cong., 1st sess., 1913. I am grateful to Randy Lawrence for alerting me to this point.

25. Ibid.; and Sheltering Arms Hospital, Case Books, item 831, located in the Archives, West Virginia University Library, Morgantown, W. Va. These hospital records reveal that a high proportion of patients were admitted with typhoid.

26. Edward E. Hunt et al., *What the Coal Commission Found* (Baltimore: Williams and Wilkins, 1925), p. 147.

27. Corbin, *Life, Work, and Rebellion*, p. 10.

28. Keith Dix, *Work Relations in the Coal Industry: The Hand-Loading Era, 1880–1930* (Morgantown, W. Va.: Institute for Labor Studies, 1977), pp. 67–68.

29. *Report of Miners' Hospital No. 3, Fairmont, W. Va.* (Charleston, W. Va.: Tribune, 1902).

30. Senate Subcommittee, *Conditions in the Paint Creek District*; and U.S. Department of the Interior, Coal Mines Administration, *A Medical Survey of the Bituminous-Coal Industry* (Washington, D.C.: U.S. Govern-

ment Printing Office, 1947).

31. Coal Mines Administration, *Medical Survey.*

32. Ibid., pp. 118, 138, 139.

33. Ibid., p. 127.

34. Ibid.

35. Senate Subcommittee, *Conditions in the Paint Creek District*, pp. 909–10.

36. Earl H. Rebhorn, "Anthraco Silicosis," *The Medical Society Reporter* (Official Journal of the Lackawanna Medical Society) 29, no. 5 (May 1935): 15.

37. David Kotelchuk, "Losing Patience: A Look Back at Corporate Medicine in the Asbestos Industry," *Health / PAC Bulletin* 2, no. 5 (May–June 1980): 1–2, 7–14.

38. See Karl Figlio, "Chlorosis and Chronic Disease in Nineteenth-Century Britain: The Social Constitution of Somatic Illness in a Capitalist Society," *International Journal of Health Services* 8, no. 4 (1978): 589–617.

39. Emery R. Hayhurst, "The Health Hazards and Mortality Statistics of Soft Coal Mining in Illinois and Ohio," *Journal of Industrial Hygiene* 1, no. 7 (November 1919): 360.

40. Ibid., pp. 366–67.

41. D. K. Brundage, E. S. Frasier and L. U. Gardner, "Exposure to Dust in Coal Mining," *The Health of Workers in Dusty Trades: III*, U.S. Public Health Service Bulletin no. 208, 1933, pp. 7–19; R. R. Sayers et al, "Anthraco-Silicosis among Hard Coal Miners," U.S. Public Health Service Bulletin no. 221, 1935; B. G. Clarke and C. E. Moffet, "Silicosis in Soft Coal Miners," *Journal of Industrial Hygiene and Toxicology* 23 (May 1941); 176–86; R. H. Flinn et al., "Soft Coal Miners Health and Working Environment," U.S. Public Health Service Bulletin no. 270, 1941; R. Harold Jones, "Pneumoconiosis Encountered in Bituminous Coal Miners," *Journal of the American Medical Association* 119, no. 8 (June 1942): 611–15. See the annotated bibliography in H. N. Doyle and T. H. Noerhren, "Pulmonary Fibrosis in Soft Coal Miners," U.S. Public Health Service Publication no. 352, 1954.

42. Leroy Gardner, as quoted in Lewis Thompson, Albert E. Russell, and J. J. Bloomfield, "General Statement and Summary of Findings," *The Health of Workers in Dustry Trades: III*, U.S. Public Health Service

Bulletin no. 208, 1933, p. 18.

43. Rebhorn, "Anthraco Silicosis," p. 17.

44. D. L. Rasmussen and C. W. Nelson, "Respiratory Function in Southern Appalachian Coal Miners," *American Review of Respiratory Disease* 103 (1971): 240–48; D. Rivers et al., "Dust Content, Radiology, and Pathology in Simple Pneumoconiosis of Coalworkers," *British Journal of Industrial Medicine* 17 (1960): 87–108; Jerome Kleinerman et al., "Pathology Standards for Coal Workers' Pneumoconiosis," *Archives of Pathology and Laboratory Medicine* 103, no. 8 (July 1979): entire special issue.

45. Corbin, *Life, Work, and Rebellion*, p. 87.

46. Ibid.; McAlister Coleman, *Men and Coal* (New York: Farrar and Rinehart, 1943).

47. Saul D. Alinsky, *John L. Lewis* (New York: Vintage Books, 1970); Keith Dix et al., *Work Stoppages and the Grievance Procedure in the Appalachian Bituminous Coal Industry* (Morgantown, W. Va.: Institute for Labor Studies, 1972), p. 5.

48. Marjorie Taubenhaus and Roy Penchansky, "The Medical Care Program of the United Mine Workers of America Welfare and Retirement Fund," in *Health Services Administration: Policy Cases and the Case Method*, ed. Roy Penchansky (Cambridge: Harvard University Press, 1968), p. 152. See also United Mine Workers of America, *Proceedings of the 39th Consecutive Constitutional Convention*, Atlantic City, N.J., 1946 (Washington, D.C.: United Mine Workers of America, 1946), 1:71.

49. Alinsky, *John L. Lewis*, p. 330.

50. G. Sinha, "Economics of Labor Welfare Funds in the Coal Industries of the United States" (Ph.D. diss., Cornell University, 1956), pp. 176–77.

51. United Mine Workers of America, Welfare and Retirement Fund, *Four-Year Summary and Review for the Year Ended June 30, 1951* (Washington, D.C.: United Mine Workers of America, 1951), pp. 23–45.

52. Curtis Seltzer, *Fire in the Hole: Miners and Managers in the American Coal Industry* (Lexington: University Press of Kentucky, 1985), pp. 61–70.

53. Sinha, "Economics of Labor Welfare Funds," pp. 187, 189.

54. Kent Barstow, "John L. Lewis versus the Doctors," *Medical Eco-*

*nomics*, November 1946, p. 56.

55. Richard Carter, *The Doctor Business* (Garden City, N.Y.: Doubleday, 1958), pp. 188, 195; see also Taubenhaus and Penchansky, "Medical Care Program of the United Mine Workers."

56. Joseph E. Martin, Jr., "Coal Miners' Pneumoconiosis," *American Journal of Public Health* 44 (May 1954): 581.

57. Lorin Kerr, "Coal Workers' Pneumoconiosis," *Industrial Medicine and Surgery* 25, no. 8 (August 1956): 355–62; Murray B. Hunter and Milton D. Levine, "Clinical Study of Pneumoconiosis of Coal Workers in Ohio River Valley," *Journal of the American Medical Association* 163, no. 1 (January 5, 1957): 1–4.

58. Author's interview with former fund staff person, Washington, D.C., December 17, 1978.

59. Murray C. Brown, "Pneumoconiosis in Bituminous Coal Miners," *Mining Congress Journal* 51, no. 8 (August 1965): 46.

60. Jan Lieben, Eugene Pendergrass, and W. Wayne McBride, "Pneumoconiosis Study in Central Pennsylvania Coal Mines," *Journal of Occupational Medicine* 3, no. 11 (November 1961): 493; and W. Wayne McBride, Eugene Pendergrass, and Jan Lieben, "Pneumoconiosis Study of Western Pennsylvania Bituminous-Coal Miners," *Journal of Occupational Medicine* 5, no. 8 (August 1963): 376.

61. Jan Lieben and W. Wayne McBride, "Pneumoconiosis in Pennsylvania's Bituminous Mining Industry," *Journal of the American Medical Association* 183, no. 3 (January 19, 1963): 176, 178; and Lieben, Pendergrass, and McBride, "Pneumoconiosis Study in Central Pennsylvania Coal Mines," p. 496.

62. W. S. Lainhart et al., *Pneumoconiosis in Appalachian Bituminous Coal Miners* (Washington, D.C.: U.S. Government Printing Office, 1969).

63. R. E. Hyatt, A. D. Kistin, and T. K. Mahan, "Respiratory Disease in Southern West Virginia Coal Miners," *American Review of Respiratory Disease* 89, no. 3 (March 1964): 387–401.

64. Ibid., p. 398.

65. Donald L. Rasmussen, M.D., "Statement," Sept. 3, 1975, photocopy of transcript in author's possession, pp. 4, 6; see also Donald L. Rasmussen et al., "Pulmonary Impairment in Southern West Virginia

Coal Miners," *American Review of Respiratory Disease* 98 (October 1968): 658–67.

66. W. Donald Ross et al., "Emotional Aspects of Respiratory Disorders among Coal Miners," *Journal of the American Medical Association* 156, no. 5 (October 1954): 484–87. It is worth noting for the historical record that, two years later, this journal refused to publish a scholarly review of British and American scientific literature on coal workers' pneumoconiosis prepared by Dr. Lorin Kerr (of the UMWA fund). Kerr's article was published in *Industrial Medicine and Surgery* (see note 57).

67. "The Coal Mine as a Tuberculosis Sanitarium," *American Labor Legislation Review* 12, no. 2 (June 1922): 131; see also Meiklejohn, "History of Lung Disease of Coal Miners in Great Britain," pt. 2, p. 94. This view apparently originated in Britain and was picked up by physicians in the United States.

## Chapter 2

1. *Webster's Seventh New Collegiate Dictionary* (Springfield, Mass.: G. and C. Merriam, 1967), p. 258.

2. This analysis draws on recent theoretical and historical literature on the labor process in the United States. The landmark work in this field is Harry Braverman's *Labor and Monopoly Capital* (New York: Monthly Review Press, 1974). See also Richard Edwards, *Contested Terrain: The Transformation of the Workplace in the Twentieth Century* (New York: Basic Books, 1979); Andrew Zimbalist, ed., *Case Studies on the Labor Process* (New York: Monthly Review Press, 1979). On the role of science and engineering in this process, see David F. Noble, *America by Design: Science, Technology, and the Rise of Corporate Capitalism* (New York: Alfred A. Knopf, 1977).

3. Carter Goodrich, *The Miner's Freedom* (Boston: Marshall Jones, 1925); John Brophy, *A Miner's Life* (Madison: University of Wisconsin Press, 1964); Keith Dix, *Work Relations in the Coal Industry: The Hand-Loading Era, 1880–1930* (Morgantown, W. Va.: Institute for Labor Studies, 1977).

4. Goodrich, *Miner's Freedom.*

5. See chap. 4, "The Work of a Miner," in Brophy's *A Miner's Life* for a thorough, first-hand account of coal mining in the hand-loading era.

6. Goodrich, *Miner's Freedom*, p. 23.

7. Brophy, *Miner's Life*, p. 45.

8. Ibid., p. 46.

9. Andrew Roy, *History of Coal Miners of the United States* (Columbus, Ohio: J. L. Trauger, 1907), p. 120: "We ask for a mouthful of fresh air amidst the mephitic blasts of death which surround us." On the legislative and collective bargaining efforts of miners and their union, see William Graebner, *Coal-Mining Safety in the Progressive Period: The Political Economy of Reform* (Lexington: University Press of Kentucky, 1976).

10. Brophy, *Miner's Life*, p. 39.

11. Henry C. Sheafer, "Hygiene of Coal-Mines," in *A Treatise on Hygiene and Public Health*, ed. Albert H. Buck (New York: William Wood, 1879), 2:244.

12. Roy, *History of Coal Miners*, pp. 449–50.

13. Morton Baratz, *The Union and the Coal Industry* (New Haven, Conn.: Yale University Press, 1955); Walter Hamilton and Helen Wright, *The Case of Bituminous Coal* (New York: Macmillan, 1925). William Graebner's "Great Expectations: The Search for Order in Bituminous Coal, 1890–1917," *Business History Review* 48, no. 1 (Spring 1974): 49–72, documents the contemporary view of the coal industry and analyzes the operators' efforts to stabilize its economic dynamics.

14. Data are from the U.S. Bureau of the Census, as reported in Joseph Gowaskie, "From Conflict to Cooperation: John Mitchell and Bituminous Coal Operators, 1898–1908," *Historian* 38, no. 4 (August 1976): 670.

15. One study actually found an inverse statistical relationship between employment levels and the rate of fatal accidents. See the discussion in Dix, *Work Relations in the Coal Industry*, pp. 101–4.

16. Roy, *History of Coal Miners*, p. 119.

17. In some cases, state law or local practice dictated that coal be shot down at the end of the day, which allowed the atmosphere to clear overnight. However, this practice was not uniform.

18. This discussion relies largely on Graebner's *Coal-Mining Safety.*

19. Ibid., p. 72.

20. Compiled by UMWA president John Mitchell in 1908 and reported by Graebner in *Coal-Mining Safety*, p. 139.

21. W. W. Keefer, as quoted in Graebner, *Coal-Mining Safety* p. 24.

22. See the discussion in chap. 3, "Safety and the Miner's Job," in Dix, *Work Relations in the Coal Industry*, pp. 67–104. Between 1906 and 1935, falls of roof and coal accounted for 53 percent of the fatalities in bituminous coal mines, followed by haulage accidents at 18 percent. Gas and dust explosions were involved in 16 percent of the occupational deaths.

23. House Committee on Education and Labor, *Legislative History of the Federal Coal Mine Health and Safety Act*, 91st Cong., 2d sess., 1970, p. 4.

24. Graebner, *Coal-Mining Safety*, p. 131.

25. Roy, *History of Coal Miners*, and Brophy, *Miner's Life*. Employers saw the issue differently. After the president of UMWA District 2 (Pennsylvania) commented on miners' respiratory disease at a legislative hearing in 1909, the first coal operator to respond stated: "I have never met a miner with asthma." As quoted in Graebner, *Coal-Mining Safety*, p. 141.

26. Graebner, *Coal-Mining Safety*, pp. 135–37, 37.

27. Ibid., p. 99.

28. *West Virginia Inspectors' Report, 1907*, p. 443, as quoted in Graebner, *Coal-Mining Safety*, p. 99.

29. John Brophy was one of the foremost proponents of nationalization. See his autobiography, *A Miner's Life*.

30. *United Mine Workers Journal*, November 26, 1914, p. 4, as quoted in Graebner, *Coal-Mining Safety*, p. 139 (hereafter cited as UMW *Journal*).

31. See David Alan Corbin, *Life, Work, and Rebellion in the Coal Fields: The Southern West Virginia Miners, 1880–1922* (Chicago: University of Illinois Press, 1981); Howard B. Lee, *Bloodletting in Appalachia* (Parsons, W. Va.: McClain, 1969); Winthrop Lane, *Civil War in West Virginia* (New York: Arno Press, 1969).

32. See McAlister Coleman, *Men and Coal* (New York: Farrar and Rinehart, 1943); Baratz, *Union and the Coal Industry*; Arthur E. Suffern, *The Coal Miners' Struggle for Industrial Status* (New York: Macmillan, 1926), p. 450.

33. This argument is presented in Graebner, "Great Expectations," and Gowaskie, "From Conflict to Cooperation."

34. "No attempt is made to make wages uniform or the earning capacity of the men equal between the different districts, or within the districts themselves, the principal object being so to regulate the scale of mining as to make cost of production practically the same in one district that it is in another, regardless of whether or not the earnings of the miners are equal." U.S. Industrial Commission, *Report of the Industrial Commission on the Relations and Conditions of Capital and Labor Employed in the Mining Industry* (Washington, D.C.: U.S. Government Printing Office, 1901) 12: 698.

35. F. L. Robbins, Chairman of the Board of the Pittsburgh Coal Company, as quoted in Gowaskie, "From Conflict to Cooperation," p. 681.

36. Coleman, *Men and Coal*, p. 122.

37. *Coal Age*, January 1949, pp. 90–93

38. See the discussion in Dix, *Work Relations in the Coal Industry*, pp. 21–29.

39. Ibid., pp. 16–29.

40. U.S. Industrial Commission, *Foreign Born in the Coal Mines* (Washington, D.C.: U.S. Government Printing Office, 1902), 15:399, as quoted in Dix, *Work Relations in the Coal Industry*, p. 19.

41. Goodrich, *Miner's Freedom*, pp. 125–30.

42. Ibid., p. 126.

43. *Keystone Coal Industry Manual* (New York: McGraw-Hill, 1973).

44. *Mines and Minerals*, December 1899, p. 205, as quoted in Dix, *Work Relations in the Coal Industry*, p. 25.

45. Dix, *Work Relations in the Coal Industry*, p. 25.

46. Author's interview with disabled coal miner, Logan County, W. Va., September 7, 1978.

47. George Korson, *Coal Dust on the Fiddle: Songs and Stories of the Bituminous Industry* (Philadelphia: University of Pennsylvania Press, 1943), p. 450.

48. George Sizemore, "Drill Man Blues," in Korson, *Coal Dust on the Fiddle*, p. 235.

49. United Mine Workers of America, *Proceedings of the 33rd Constitu-*

*tional Convention* (Indianapolis: United Mine Workers of America, 1934), p. 192; see pages 187–206 for a fascinating debate on the introduction of machinery. At this time, opposition to mechanization represented in part the voice of the older craft miner, who was being forced to change drastically his customary work practices. Opposition also arose out of the special problems accompanying the transition from hand-loading to mechanization (for example, exhausting pressure on hand loaders, who were, in many cases, older men, to keep pace with the machines; new health and safety problems; and a work force divided between machine operators on a daily wage and hand loaders paid by the ton or car).

50. *Keystone Coal Mine Directory* (New York: McGraw-Hill, 1977). See Michael Yarrow, "The Labor Process in Coal Mining: The Struggle for Control," in Zimbalist, *Case Studies*, pp. 170–92.

51. Baratz, *Union and the Coal Industry*, pp. 125–27.

52. In 1954, representatives from the forty largest commercial coal companies in the United States, who together controlled almost half of the country's annual production, met on three separate occasions. The group selected an eleven-member subcommittee to "spell out consolidation methods that could be used by any group of coal producers" (*Business Week*, July 10, 1954, p. 31). See the discussion in Curtis Seltzer, *Fire in the Hole: Miners and Managers in the American Coal Industry* (Lexington: University Press of Kentucky, 1985), p. 74.

53. Although very different, the two best biographies of Lewis are Saul D. Alinsky, *John L. Lewis* (New York: Vintage Books, 1970), and Melvyn Dubofsky and Warren Van Tine, *John L. Lewis* (New York: Quadrangle, 1977).

54. J.B.S. Hardman, "John L. Lewis, Labor Leader and Man: An Interpretation," *Labor History* 2, no. 1 (Winter 1961): 15.

55. Bernard Karsh and Jack London, "The Coal Miners: A Study of Union Control," *Quarterly Journal of Economics* 68, no. 3 (August 1954): 421, quoting a miner.

56. See Dubofsky and Van Tine, *John L. Lewis*, and Alinsky, *John L. Lewis*.

57. See John L. Lewis, *The Miners' Fight for American Standards* (Indianapolis: Bell, 1925).

58. Coleman, *Men and Coal*, pp. 126–35.

59. *Coal Age*, February 1951, p. 139.

60. "Uniting for Strength," *Nation's Business*, January 1967, p. 49.

61. Dubofsky and Van Tine, *John L. Lewis*, p. 496.

62. See Tom Bethell, "Conspiracy in Coal," in *Appalachia in the Sixties: Decade of Reawakening*, ed. David S. Walls and John B. Stephenson (Lexington: University Press of Kentucky, 1972), pp. 81–83.

63. See chap. 5, "The Lewis-Love Axis and the Great Shakeout," in Seltzer, *Fire in the Hole*, pp. 61–70.

64. *New York Times*, March 6, 1970, p. 12.

65. Ibid. See also John Peter David, "Earnings, Health, Safety, and Welfare of Bituminous Coal Miners since the Encouragement of Mechanization by the United Mine Workers of America" (Ph.D. diss., West Virginia University, 1972).

66. *1947 Bituminous Coal Wage Agreement*, p. 1.

67. For example, from the *New York Times* (May 10, 1954, p. 1): "One of the sickest of the country's industries [coal] is getting sicker." Throughout the early 1950s, editorials in the industry magazine, *Coal Age*, reflected a concern for negating coal's "sick" image.

68. United Mine Workers of America, *Officers' Report to the 41st Consecutive Constitutional Convention* (Washington, D.C.: United Mine Workers of America, 1952), p. 111. Emphasis added.

69. See Senate Committee on Labor and Public Welfare, Subcommittee on Labor, *To Amend the Federal Coal Mine Safety Act*, Hearings, 85th Cong., 2d sess., 1958.

70. As quoted in Curtis Seltzer, "The United Mine Workers of America and the Coal Operators: The Political Economy of Coal in Appalachia, 1950–1973" (Ph.D. diss., Columbia University, 1977), p. 575.

71. Senate Subcommittee on Labor, *To Amend the Federal Coal Mine Safety Act*, p. 126.

72. As quoted in Seltzer, "UMWA and the Coal Operators," p. 576.

73. *Coal Age*, March 1953, p. 136.

74. *New York Times*, March 28, 1954, p. 82.

75. Joseph Finley, *The Corrupt Kingdom* (New York: Simon and Schuster, 1972), pp. 191–93.

76. Author's interview with former fund staff person, Washington, D.C., March 15, 1978.

77. UMW *Journal*, May 1, 1971, p. 2.

78. See the discussion in Seltzer, *Fire in the Hole*, pp. 82–83; for more recent developments, see pp. 178–85.

79. Finley, *Corrupt Kingdom*, p. 200.

80. Ibid., p. 167.

81. *Wall Street Journal*, December 26, 1968, p. 1.

82. Ibid.

83. See *People's Appalachia* 2, no. 3 (July 1972), special issue on urban migrants; Abt Associates, *The Causes of Rural to Urban Migration among the Poor*, Special Report to the U.S. Office of Economic Opportunity (March 1970), especially chap. 17; and, for an excellent fictional account of the experiences of an Appalachian migrant, Harriette Arnow, *The Dollmaker* (New York: Macmillan, 1954).

84. I am grateful to Dr. Jerome Pickard, demographer for the Appalachian Regional Commission, for making available to me a wealth of unpublished data on migration. This figure is derived from census records.

85. For example, between 1950 and 1970, the number of black miners declined by 87 percent, from 5,150 to 678 in McDowell County, West Virginia. This reduction was more drastic than white miners' 63 percent decline. U.S. Department of Commerce, Bureau of the Census, *1950 Census of the Population*, vol. 2, pt. 48; and *1970 Census of the Population*, vol. 1, pt. 50. Interviews with both black and white miners confirmed the role of racial discrimination in the layoffs.

86. Author's interview with disabled coal miner, Logan County, W. Va., September 6, 1978.

87. *Keystone Coal Industry Manual*, p. 430.

88. Author's interview with former coal miner, Kanawha County, W. Va., October 2, 1978. Emphasis added.

89. United Mine Workers of America, *Proceedings of the 42nd Constitutional Convention* (Washington, D.C.: United Mine Workers of America, 1956), p. 325.

90. U.S. Department of Labor, Mine Safety and Health Administration, *Injury Experience in Coal Mining, 1977* (Washington, D.C.: U.S.

Government Printing Office, 1979), p. 131; National Coal Association, *Coal Data 1976* (Washington, D.C.:? National Coal Association, 1977); and U.S. Congress, Office of Technology Assessment, *The Direct Use of Coal* (Washington, D.C.: U.S. Government Printing Office, 1979), pp. 276, 278. Care must be exercised when analyzing mine safety data to distinguish between underground and surface mining. The rise in surface production and workers has generated a decline in aggregate fatality and injury rates; the aggregate rates, however, do not reveal the frequency of death and injury underground.

91. Author's interview with disabled coal miner, Logan County, W. Va., September 7, 1978.

92. Author's interview with disabled coal miner, Mingo County, W. Va., September 12, 1978.

93. United Mine Workers of America, *Proceedings of the 42nd Convention*, p. 322.

94. Ibid., pp. 326–27.

95. See *War in the Coal Fields* (1931, reprint, Huntington, W. Va.: Appalachian Movement Press, 1972); *The West Virginia Miners' Union, 1931* (1931, reprint, Huntington, W. Va.: Appalachian Movement Press, 1972).

96. See U.S. Department of Labor, Bureau of Labor Statistics, *Work Stoppages: Bituminous-Coal Mining Industry*, BLS Report no. 95 (Washington, D.C.: U.S. Government Printing Office, 1955); U.S. Department of Labor, Bureau of Labor Statistics, *Collective Bargaining in the Bituminous Coal Industry*, BLS Report no. 514 (Washington, D.C.: U.S. Government Printing Office, 1977); Keith Dix et al., *Work Stoppages and the Grievance Procedcure in the Appalachian Bituminous Coal Industry* (Morgantown, W. Va.: Institute for Labor Studies, 1972); United Mine Workers of America, *Proceedings of the 42nd Convention*, pp. 404–5; and *New York Times*, July 24, 1955, p. 51. For example, after repeated wildcats at the Robena mine, Lewis informed the local president that his suspension from the union had been recommended. The outcome of this dispute was not recorded.

97. United Mine Workers of America, *Proceedings of the 42nd Convention*, p. 312.

## Chapter 3

1. Mainstream social movement theory has conservative origins in the study of "collective behavior," which emphasizes the irrational and destructive aspects of strikes, riots, and other collective protests. The class fear and bias of the privileged in the wake of the French Revolution are evident in this tradition. See Edmund Burke, *Reflections on the Revolution in France* (Indianapolis: Bobbs-Merrill, 1955); and Gustave Le Bon, *Psychologie des Foules* (Paris: Alcan, 1895). Adherents to this school of thought frequently used "contagion" as a metaphor for the supposedly arbitrary and irrational spread of collective behavior. The title of this chapter reappropriates this term to indicate just the opposite: the contagions of disease and rebellion both have social origins that can by analyzed.

More recent theorists tend to disavow the biases of this tradition, though they remain evident in the influential works of sociologists like Talcott Parsons and Neil Smelser, who view collective protest as deviant and nonrational. See Parsons's *The Social System* (New York: Free Press, 1951), Smelser's *Theory of Collective Behavior* (New York: Free Press, 1962), and the discussion of these and other theorists in Anthony Oberschall's *Social Conflict and Social Movements* (Englewood Cliffs, N.J.: Prentice-Hall, 1973), pp. 1–29.

2. Emiseration as a source of social protest is commonly attributed to Karl Marx, though social movement theorists usually abstract this single point from his historical analysis of structural contradictions in capitalism, and place it alongside other causes in a generalized, transhistorical canon. Some theorists emphasize "rising expectations" as a source of protest, while others stress the breakdown of daily routines and structures that ordinarily regulate behavior. See the summaries in Roberta Ash, *Social Movements in America* (Chicago: Markham, 1972); Michael Useem, *Protest Movements in America* (Indianapolis: Bobbs-Merrill, 1975); and Oberschall, *Social Conflict and Social Movements.*

3. For a discussion of how acquiescence is maintained and rebellion generated in an Appalachian context, see John Gaventa, *Power and Powerlessness: Quiescence and Rebellion in an Appalachian Valley* (Chicago: University of Illinois Press, 1980).

4. Author's interview with disabled miner, Fayette County, W. Va., August 28, 1978.

5. Widows under fifty years of age without dependents were ineligible.

6. Widows with dependent children were sometimes eligible for Aid to Families with Dependent Children. In 1956, permanently and totally disabled workers became eligible for social security retirement benefits at age fifty, and in 1960, the age requirement was dropped for those who were otherwise eligible. Disabled workers were also eligible for workers' compensation, but the often insurmountable obstacle to obtaining benefits under either program was the need to prove "permanent and total disability" and, in the case of workers' compensation, the need to prove that disability was work related.

7. Only disabled miners receiving workers' compensation for their injury were eligible for continuing assistance up to four years. Many other disabled miners lost their hospital cards immediately, as the fund trustees ruled ineligible miners who for the past year had been unemployed or had worked in a nonunion mine. See United Mine Workers of America, Welfare and Retirement Fund, *Report for the Year Ending June 30, 1961* (Washington, D.C.: Welfare and Retirement Fund, 1961), p. 16.

8. Author's interview with activist in the Association of Disabled Miners and Widows (and wife of a disabled miner), Raleigh County, W. Va., September 22, 1978.

9. Brit Hume, *Death and the Mines* (New York: Grossman, 1971), p. 43.

10. Jeanne Rasmussen, "On the Outside Looking In," *Mountain Life and Work*, September 1969, p. 9. On the fund's pension and other benefits policies, see Joseph Finley, *The Corrupt Kingdom* (New York: Simon and Schuster, 1972), pp. 178–204; and Hume, *Death and the Mines*, pp. 29–52.

11. Beckley (W. Va.) *Post Herald*, July 16, 1960, p. 1.

12. *Raleigh Register* (Raleigh County, W. Va.), August 8, 1960, p. 12. This account of the disabled miners' and widows' organizations is based on information from interviews and news coverage in the *Raleigh Register* and Beckley *Post Herald*.

13. *Raleigh Register*, May 1, 1967, p. 3. The earliest news account of

the disabled miners and widows I could find was on this date. However, the generally accepted year of the organization's inception, cited by members and supporters in several interviews, is 1966.

14. *Wall Street Journal*, December 26, 1968, p. 1, quoting a male member of the organization.

15. See Rasmussen, "On the Outside Looking In."

16. They demanded a longer statute of limitations for silicosis claims. *Raleigh Register* and Beckley *Post Herald*, January 13, 1968.

17. Attorneys for the plaintiffs were Harry Huge, from the firm of Arnold and Porter; Paul Kaufman, from Charleston, West Virginia; and Harry Caudill, author of *Night Comes to the Cumberlands*, of Whitesburg, Kentucky. Huge was later appointed a trustee of the fund and presided over its virtual demise in 1978.

18. Rasmussen, "On the Outside Looking In," p. 21.

19. On the Blankenship case, see Finley, *Corrupt Kingdom*, pp. 199–200. For the union's viewpoint, see the UMW *Journal*, March 1 and May 1, 1971.

20. Author's interview with a member of the Association of Disabled Miners and Widows, Raleigh County, W. Va., September 22, 1978.

21. A candid summary of the major compromises and controversies surrounding the war on poverty may be found in White House Task Force, *The Office of Economic Opportunity during the Administration of President Lyndon B. Johnson* (Washington, D.C.), vol. 1. For a thoroughly cynical, anecdotal view of the Johnson administration, including the poverty program, see Robert Sherrill, *The Accidental President* (New York: Grossman, 1967). A more analytical, radical critique of the war on poverty may be found in Frances Fox Piven and Richard A. Cloward, *Regulating the Poor* (New York: Random House, 1971). A primary source for the bitter political conflicts surrounding certain programs is Senate Committee on Labor and Public Welfare, Subcommittee on Employment, Manpower, and Poverty, *Examination of the War On Poverty*, Hearings, 90th Cong., 1st sess., 1967.

22. This view is not shared by most analysts. The classic studies of the origins of the war on poverty are Peter Marris and Martin Rein, *Dilemmas of Social Reform* (Chicago: Routledge and Kegan Paul, 1967); James L. Sundquist, *Politics and Policy: The Eisenhower, Kennedy, and Johnson Years* (Washington, D.C.: Brookings Institution, 1968); and Sar A.

Levitan, *The Great Society's Poor Law* (Baltimore: Johns Hopkins University Press, 1969).

23. See President Lyndon B. Johnson's first State of the Union message, *Congressional Quarterly Almanac*, vol. 20 (Washington, D.C.: Congressional Quarterly Service, 1965), pp. 862–64.

24. *Economic Report of the President*, 88th Cong., 2d sess., 1964, p. 66.

25. Christopher Jencks may have been the first to coin this phrase. See his "Johnson vs. Poverty," *New Republic*, March 28, 1964, p. 18.

26. Sherrill, *Accidental President*, p. 176.

27. Saul Alinsky was one of the most visible leftist critics of the war on poverty. See "A Professional Radical Moves in on Rochester: Conversations with Saul Alinsky," *Harper's*, June, July, 1965. For a glimpse of the view from the Right, see Senate Subcommittee on Employment, Manpower, and Poverty, *Examination of the War on Poverty.*

28. *Congressional Quarterly Almanac*, vol. 17 (Washington, D.C.: Congressional Quarterly Service, 1962), p. 249.

29. See President's Appalachian Regional Commission, *Appalachia* (Washington, D.C.: U.S. Government Printing Office, 1964).

30. *Congressional Quarterly Almanac*, vol. 21 (Washington, D.C.: Congressional Quarterly Service, 1966), pp. 788–97.

31. "No analysis of the regional problem has failed to identify the historic and persisting barrier-effect of its mountain-chains as a primary factor in Appalachian underdevelopment." President's Appalachian Regional Commission, *Appalachia*, p. 32.

32. See David S. Walls and John B. Stephenson, eds., *Appalachia in the Sixties: Decade of Reawakening* (Lexington: University Press of Kentucky, 1972); Bruce Jackson, "In the Land of the Shadows: Kentucky," in *How We Lost the War on Poverty*, ed. Marc Pilisuk and Phyllis Pilisuk (New Brunswick, N.J.: Transaction Books, 1973), pp. 78–103.

33. See Robert Coles and Joseph Brenner, "American Youth in a Social Struggle: The Appalachian Volunteers," *American Journal of Orthopsychiatry* 38, no. 1 (January 1968): 31–46.

34. See Huey Perry, *"They'll Cut Off Your Project": A Mingo County Chronicle* (New York: Praeger, 1972); Walls and Stephenson, *Appalachia in the Sixties*; and Richard Harris, *Freedom Spent* (Boston: Little, Brown, 1974), pp. 123–312.

35. Congressional debate over the Green amendment was pro-

tracted and intense. See *Congressional Record, House*, vol. 113, pt. 24, especially November 14, 1967.

36. Author's interview with former Appalachian Volunteer and community action staff person, Raleigh County, W. Va., September 18, 1978.

37. Author's interview with former Appalachian Volunteers staff person, Kanawha County, W. Va., October 4, 1978.

38. The only complimentary account of Boyle's leadership of the UMWA may be found in the pages of the UMW *Journal*, 1963–72. For the other side of the story, see Finley, *Corrupt Kingdom*, and Hume, *Death and the Mines*.

39. See John Peter David, "Earnings, Health, Safety, and Welfare of Bituminous Coal Miners since the Encouragement of Mechanization by the United Mine Workers of America" (Ph.D. diss., West Virginia University, 1972). David demonstrates that coal miners lost ground in wages, benefits, and working conditions during this period as compared with unionized workers in other basic industries.

40. *Wall Street Journal*, December 18, 1963, p. 3.

41. Ibid., September 3, 1963, p. 9.

42. Ibid., December 18, 1963, p. 3.

43. Ibid., February 7, 1964, p. 1.

44. Ibid.

45. Ibid., March 24, 1964, p. 2.

46. *1964 Bituminous Coal Wage Agreement*, as reproduced in United Mine Workers of America, *Proceedings of the 44th Constitutional Convention* (Washington, D.C.: United Mine Workers of America, 1964), 1:46.

47. Ibid.: "The right to install and operate new types of equipment is recognized.

48. *New York Times*, April 3, 1964, p. 18; *Wall Street Journal*, April 2, 1964, p. 4.

49. For a record of strike activity, see U.S. Department of Labor, Bureau of Labor Statistics, *Collective Bargaining in the Bituminous Coal Industry*, *BLS* Report no. 514 (Washington, D.C.: U.S. Government Printing Office, 1977), p. 5.

50. This was of more than symbolic importance. When conventions were held in locations distant from the coalfields, large locals sometimes were unable to finance their full complement of delegates.

51. See Hume, *Death and the Mines*, p. 47.

52. Ibid., pp. 43–52.

53. United Mine Workers of America, *Proceedings of the 44th Convention*, 1:411.

54. In what was clearly a centrally coordinated effort (although not necessarily by Boyle in Washington, D.C.), several locals from District 19 submitted resolutions demanding that certain dissident leaders, including Steve Kochis, be suspended from the union and barred from holding office. The resolutions were referred to the Constitution Committee. See United Mine Workers of America, *Proceedings of the 44th Convention*, 2:59.

55. Hume, *Death and the Mines*, p. 52.

56. U.S. Department of Labor, BLS, *Collective Bargaining in the Bituminous Coal Industry*, pp. 8–9.

57. *Wall Street Journal*, September 24, 1965, p. 4.

58. *Business Week*, August 15, 1964, p. 70.

59. *Wall Street Journal*, September 24, 1965, p. 4.

60. *New York Times*, April 12, 1966, p. 1.

61. See *Business Week*, May 7, 1966, p. 122.

62. U.S. Department of Labor, BLS, *Collective Bargaining in the Bituminous Coal Industry*, p. 5.

63. See *Coal Age*, January 1965, pp. 62–63; *Business Week*, September 16, 1967, pp. 164–67; *Wall Street Journal*, September 6, 1967, p. 1; and *New York Times*, October 14, 1967, p. 7.

64. By 1964, the median age of the UMWA work force had risen to slightly above forty-eight years. United Mine Workers of America, *Officers' Report to the 47th Constitutional Convention* (Washington, D.C.: United Mine Workers of America, 1976), p. 3.

## Chapter 4

1. This description of the Farmington disaster (also known as the Mannington disaster, after another nearby town) is based primarily on newspaper accounts from the *New York Times, Washington Post, Wall Street Journal*, and (Charleston) *Gazette*.

2. UMW *Journal*, December 1, 1968, p. 4.

3. Farmington was by far the worst mine disaster since the television had become a common household item.

4. *Washington Post*, November 23, 1968, p. 3.

5. See the *New York Times*, November 21, 1968, p. 1; November 22, 1968, p. 78; and November 24, 1968, section 4, p. 6.

6. *Washington Post*, November 22, 1968, p. 1; *Wall Street Journal*, December 3, 1968, p. 1; and *New York Times*, March 30, 1968, section 6, p. 25.

7. As quoted in Brit Hume, *Death and the Mines* (New York: Grossman, 1971), p. 9.

8. UMW *Journal*, December 1, 1968, pp. 3–4.

9. *New York Times*, November 22, 1968, p. 50.

10. UMW *Journal*, December 1, 1968, p. 11.

11. See Ben Franklin, "The Scandal of Death and Injury in the Mines," in *Appalachia in the Sixties: Decade of Reawakening* ed. David S. Walls and John B. Stephenson (Lexington: University Press of Kentucky, 1972), pp. 92–108; Davitt McAteer, *Coal Mine Health and Safety: The Case of West Virginia* (New York: Praeger, 1970); Hume, *Death and the Mines*; Robert Sherrill, "The Black Lung Rebellion," *Nation*, April 28, 1969, p. 531; and Bill Montgomery, "The Black-Lung Battle," *Mountain Life and Work*, November 1968, p. 9. Some of these authors emphasized the role of professionals like Ralph Nader in "starting" the black lung movement. Sherrill, e.g., observed that miners were "responding to an anger that had been deliberately drummed into them over a period of several months by nonunion outsiders." Such observers seemed to confuse their own discovery of black lung with miners' recognition of the problem. Ironically, they echoed the charges of UMWA leaders, who asserted that the movement was entirely the work of "outside rabble rousers."

12. The mine explosion occurred in northern West Virginia, an area distinct in its economy, labor history, and union politics from the southern region. This partly explains the limited impact of the disaster on miners in southern West Virginia.

13. United Mine Workers of America, *Proceedings of the 40th Constitutional Convention* (2 vols.; Washington, D.C.: United Mine Workers of America, 1948), 1:327.

14. See Pat Forman, "Scandal at Gauley Bridge," *Health / PAC Bulle-*

*tin*, no. 79 (November–December 1977), pp. 9–16.

15. See the *Proceedings* of the UMWA conventions for these years, especially the supplementary volumes containing all resolutions.

16. By 1969, Alabama, Virginia, and Pennsylvania recognized coal workers' pneumoconiosis as compensable. The Pennsylvania law was due in part to the lobbying of Jock Yablonski, president of UMWA District 5 in the western part of the state. See Peter S. Barth with H. Allan Hunt, *Workers' Compensation and Work-Related Illnesses and Diseases* (Cambridge: MIT Press, 1980), for a detailed discussion of current problems in obtaining workers' compensation for occupational disease.

17. This practice was mentioned by several miners during interviews with the author. Aggravating the situation for miners in southern West Virginia was an informal agreement between UMWA district officials and local coal operators (the so-called Olga Agreement, for Olga Coal Company); this pledged the union not to contest discharges of miners who had received compensation awards for more than 10 percent disability due to occupational lung disease. Author's interview with former miner, Kanawha County, W. Va., October 2, 1978.

18. See the National Conference on Medicine and the Federal Coal Mine Health and Safety Act of 1969, *Papers and Proceedings*, Washington, D.C., June 15–18, 1970.

19. See *Raleigh Register* (Raleigh County, W. Va.), September 3, 1964, p. 1; Beckley (W. Va.) *Post Herald*, May 13, 1965, p. 1.

20. Author's interview with miner's wife and black lung activist, Kanawha County, W. Va., October 4, 1978.

21. Author's interview with disabled miner, Raleigh County, W. Va., July 24, 1978.

22. See the United Mine Workers of America, *Proceedings of the 45th Constitutional Convention* (2 vols.; Washington, D.C.: United Mine Workers of America, 1968).

23. Ibid., pp. 320–21.

24. Author's interview with black lung activist, Raleigh County, W. Va., September 26, 1978.

25. Author's interview with disabled miner, Fayette County, W. Va., August 23, 1978.

26. Ibid.

27. Author's interview with black lung activist, Raleigh County, W. Va., July 13, 1978.

28. Author's interview with disabled miner, Kanawha County, W. Va., September 30, 1978.

29. *Raleigh Register*, December 23, 1968, p. 2.

30. Author's interview with disabled miner, Kanawha County, W. Va., September 30, 1978.

31. See Hume, *Death and the Mines*. Interviews with participants confirmed most of the details of Hume's account.

32. *Gazette-Mail* (Charleston, W. Va.,), January 12 and 20, 1968.

33. Author's interview with black lung activist, Raleigh County, W. Va., September 26, 1978.

34. Author's interview with disabled miner's wife and black lung activist, Fayette County, W. Va., July 26, 1978.

35. Author's interview with black lung activist and miner's wife, Kanawha County, W. Va., October 4, 1978.

36. Ibid.

37. "Your local union has no authority to donate money from the treasury, to some unknown group which, in my opinion, is dual to the UMWA, to be used for any purpose they see fit." This message went out in a letter to all local unions in District 17 (southern West Virginia) and was signed by R. R. Humphreys, district president. (As quoted in Hume, *Death and the Mines*, p. 122.)

38. See Charleston (W. Va.) *Daily Mail*, January 27, 1969, p. 1; Huntington (W. Va.) *Herald-Dispatch*, January 27, 1969, pp. 1–2; Charleston (W. Va.) *Gazette*, January 27, 1969, p. 1.

39. Author's interview with disabled miner's wife and black lung activist, Kanawha County, W. Va., September 30, 1978.

40. Charleston *Gazette*, February 12, 1969, p. 1.

41. Beckley *Post Herald*, January 15, 1969, p. 1.

42. Hume, *Death and the Mines*, p. 126.

43. *Raleigh Register*, February 18, 1969, p. 1.

44. Author's interview with working miner, Kanawha County, W. Va., September 29, 1978.

45. Author's interview with disabled miner, Mingo County, W. Va., September 12, 1978.

46. Author's interview with disabled miner, Kanawha County, W.

Va., September 30, 1978.

47. Beckley *Post Herald*, February 24, 1969, p. 1.

48. As quoted in Hume, *Death and the Mines*, p. 139.

49. Ibid., p. 122.

50. Charleston *Daily Mail*, February 27, 1969, p. 1. The operators lost this suit.

51. Charleston *Gazette*, February 10, 1969, p. 1.

52. Charleston *Daily Mail*, January 15, 1969.

53. Ibid., January 24, 1969; and Charleston *Gazette*, January 25, 1969.

54. Author's interview with West Virginia state legislator, September 20, 1978.

55. West Virginia state legislator, as quoted in Bennett M. Judkins, "The Black Lung Association: A Case Study of a Modern Social Movement" (Ph.D. diss., University of Tennessee, 1975), p. 212.

56. *Washington Post*, February 27, 1969, p. 4.

57. Author's interview with disabled miner, Kanawha County, W. Va., September 30, 1978.

58. Charleston *Daily Mail*, February 28, 1969, p. 1.

59. Ibid., March 1, 1969, p. 1.

60. *Raleigh Register and Post Herald*, March 1, 1969, pp. 1, 8.

61. This analysis is based on copies of the relevant bills, which were retrieved for me nine years later by the staff of the clerk's offices in the West Virginia House of Delegates and the West Virginia Senate. I am grateful for their efforts.

62. Delegate Cleo Jones (R–Kanawha County), as quoted in Hume, *Death and the Mines*, p. 147.

63. *Wall Street Journal*, March 7, 1969, p. 2.

64. *Acts of the West Virginia Legislature*, regular sess., 1969, 2d extraordinary session, 1968, pp. 1267–96.

65. Charleston *Gazette-Mail*, March 9, 1969, p. 1.

66. Author's interview with black lung activist, Raleigh County, W. Va., September 26, 1978.

67. Author's interview with black lung activist, Raleigh County, W. Va., July 13, 1978.

68. Edward Wieck, *Preventing Fatal Explosions in Coal Mines* (New

York: Russell Sage Foundation, 1942), p. 131.

69. James Ridgeway, *The Last Play* New York: E. P. Dutton, 1973), p. 167.

70. *Business Week*, October 16, 1965, p. 166.

71. Curtis Seltzer, "The United Mine Workers of America and the Coal Operators: The Political Economy of Coal in Appalachia, 1950–1973" (Ph.D. diss., Columbia University, 1977), pp. 815, 832.

72. As quoted in Thomas O'Hanlon, "Anarchy Threatens the Kingdom of Coal," *Fortune*, January 1971, p. 81.

73. Ibid.; and *Coal Age*, May 1967, pp. 60–66; February, 1968, p. 56. John Corcoran, president of Consol, estimated that the industry would need thirty thousand new workers over the next ten years (1969–79). Senate Committee on Labor and Public Welfare, Subcommittee on Labor, *Coal Mine Health and Safety*, Hearings, 91st Cong., 1st sess., 1969, pt. 2, p. 552.

74. *Congressional Record, House*, June 9, 1969, p. 15045. Yablonski's entire platform and announcement speech were placed in the *Record* by Representative Ken Hechler (W. Va.).

75. See Senate Committee on Labor and Public Welfare, Subcommittee on Labor, *United Mine Workers Election*, Hearings, 91st Cong., 2d sess., 1970, pt. 1, especially pp. 475–76, 489–90.

76. See the analysis in O'Hanlon, "Anarchy Threatens the Kingdom of Coal," p. 81.

77. Ibid., p. 151.

78. Interview with Gary Sellers by Thomas N. Bethell, xeroxed transcript.

79. Senate Subcommittee on Labor, *Coal Mine Health and Safety*, 1969, pt. 2, pp. 551, 556, 568.

80. P.L. 91-173, "Federal Coal Mine Health and Safety Act of 1969," sec. 4.

81. *Wall Street Journal*, September 19, 1969, p. 8; *New York Times*, September 21, 1969, p. 66.

82. See U.S. Department of Labor, Mine Safety and Health Administration, *Injury Experience in Coal Mining, 1977*, (Washington, D.C.: U.S. Government Printing Office, 1979), pp. 129–41. The rate of disabling injuries underground actually increased after passage of the law.

83. P.L. 91-173, "Federal Coal Mine Health and Safety Act of 1969," Title IV, "Black Lung Benefits," sec. 401–26.

84. See House Committee on Education and Labor, *Legislative History of the Federal Coal Mine Health and Safety Act*, 91st Cong., 2d sess., 1970, pp. 1085–95.

85. P.L. 91-173, "Federal Coal Mine Health and Safety Act of 1969," Title II, "Interim Mandatory Health Standards," sec. 201–6.

86. *New York Times*, December 30, 1969, p. 1.

87. See Trevor Armbrister, *Act of Vengeance* (New York: E. P. Dutton, 1975).

88. These responses represent the two ends of a continuum, "a range from repression to facilitation," and they are not mutually exclusive. See Anthony Oberschall, *Social Conflict and Social Movements* (Englewood Cliffs, N.J.: Prentice-Hall, 1973); Charles Tilly, *From Mobilization to Revolution* (Reading, Mass.: Addison-Wesley, 1978); and Roberta Ash, *Social Movements in America* (Chicago: Markham, 1972) for overviews. A central concern of Charles Tilly is to explain the changing *forms* of protest, and he includes the influence of the state in his analysis. See *From Mobilization to Revolution*, especially pp. 143–88, and the more recent work, which he co-edited with Louise A. Tilly, *Class Conflict and Collective Action* (Beverly Hills, Calif.: Sage Publications, 1981). Frances Fox Piven and Richard A. Cloward are also concerned to explain the influence of the state on the timing, form, and impact of protest. See *Poor People's Movements: Why They Succeed, How They Fail* (New York: Pantheon Books, 1977).

No theorists of social movements have explored the state's role in shaping the actual goals and content of protest; however, for scholars interested in political power and social control, this is a central issue. For a sampling of different approaches, see Ralf Dahrendorf, *Class and Class Conflict in Industrial Society* (Stanford, Calif.: Stanford University Press, 1959), especially on "conflict regulation"; Steven Lukes, *Power: A Radical View* (London: Macmillan, 1974), on the forms and mechanisms of power; and Nicos Poulantzas, *Political Power and Social Classes* (London: Verso Editions/ NLB, 1978), on the relationship between dominated classes and the capitalist state.

89. Paul F. Clark, *The Miners' Fight for Democracy: Arnold Miller and the*

*Reform of the United Mine Workers* (Ithaca: Cornell University Press, 1981), p. 24.

90. See James Weinstein, *The Corporate Ideal in the Liberal State* (Boston: Beacon Press, 1968); Dan Berman, *Death on the Job* (New York: Monthly Review Press, 1978); and Barth with Hunt, *Workers' Compensation and Work-Related Illnesses and Diseases.*

91. *Acts of the West Virginia Legislature,* 1969, sec. 23-4-6(e), p. 1278.

92. See Barth with Hunt, *Workers' Compensation and Work-Related Illnesses and Diseases*, concerning problems for claimants in obtaining compensation, especially for occupational disease.

93. U.S. Department of Labor, *An Interim Report to Congress on Occupational Diseases* (Washington, D.C.: U.S. Government Printing Office, 1980), p. 60.

94. Barth with Hunt, *Workers' Compensation and Work-Related Illnesses and Diseases*, p. 125.

95. See Berman, *Death on the Job*, table 5, pp. 242–43.

96. West Virginia Workmen's Compensation Fund, *Annual Report and Financial Statement*, year ending June 30, 1970, table 39-A, p. 57. Thirty-nine claims were awarded during the program's first year; twenty-eight were temporary benefits for partial disability. Following establishment of the federal black lung benefits program, eligibility for workers' compensation due to occupational pneumoconiosis began to loosen up slightly in West Virginia; however, it remains extremely difficult to win a lifetime award for permanent and total disability. For example, in the three consecutive fiscal years ending on June 30, 1978, 5,147 coal miners were pronounced disabled due to occupational pneumoconiosis by the West Virginia workers' compensation system; 92 percent of them, however, received only temporary benefits, in most cases lasting for less than two years. During the same three-year period, sixty-one widows received compensation. See West Virginia Workmen's Compensation Fund, *Annual Report and Financial Statement*, year ending June 30, 1976; year ending June 30, 1977; and year ending June 30, 1978.

97. *Acts of the West Virginia Legislature,* 1969, sec. 23-4-8a, p. 1284.

98. Senate Subcommittee on Labor, *Coal Mine Health and Safety*, 1969, pt. 1, pp. 463–64.

99. See U.S. Department of Labor, MSHA, *Injury Experience in Coal Mining, 1977*; and the 1978, 1979, 1980, 1981, and 1982 editions of the same report. In 1978, MSHA changed certain reporting and processing protocols on safety statistics and began expressing injury and fatality rates as the number of incidents per 200,000 employee-hours (as opposed to 1,000,000 employee-hours, used previously). The comparability of statistics prior to 1977 and those from more recent years is therefore reduced. However, even when more recent years are excluded from consideration, the impact of the 1969 act on disabling injuries underground remains questionable. In 1971 and 1972, the rates exceeded those of the preceding twenty years. (See the calculations in Congress of the United States, Office of Technology Assessment, *The Direct Use of Coal* (Washington, D.C.: U.S. Government Printing Office, 1979), p. 278.) Aggregate data for all miners suggest a more favorable picture, but they are misleading because of the inclusion of surface miners, whose injury rates are lower than deep miners', and whose relative numbers have been growing since World War II.

100. U.S. Comptroller General, *Improvements Still Needed in Coal Mine Dust-Sampling Program and Penalty Assessments and Collections* (Washington, D.C.: U.S. Government Printing Office, 1975), p. 15.

101. Internal memorandum from Welby G. Courtney, research supervisor, U.S. Bureau of Mines, Dust Control Group, Pittsburgh, Pennsylvania, to Thomas V. Falkie, director, U.S. Bureau of Mines, Washington, D.C., November 29, 1974. As quoted in UMW *Journal*, January 16–31, 1976, p. 6.

102. U.S. Comptroller General, *Improvements Still Needed in Coal Mine Dust-Sampling Program*, p. 15.

103. Ibid.

104. See *New York Times*, March 26 and April 9, 1979, and Charleston *Gazette*, October 15, November 11, and December 5, 1980.

105. West Virginia Workmen's Compensation Fund, *Annual Report and Financial Statement*, year ending June 30, 1970, p. 84.

## Chapter 5

1. This estimate was put forth in testimony before the House Subcommittee on Labor during congressional debate over the black lung

compensation enabling legislation. It was based primarily on the U.S Public Health Service's study of CWP among miners in Pennsylvania. The estimate referred only to those "afflicted" with disease, not those who were totally disabled by it, which presumably would have been a lower number. See House Committee on Education and Labor, *Legislative History of the Federal Coal Mine Health and Safety Act*, 91st Cong., 2d sess., 1970, p. 15.

2. Data were provided by the U.S. Department of Health and Human Services, Social Security Administration, Bureau of Disability Insurance. See the *Annual Report on Part B of Title IV of the Federal Coal Mine Health and Safety Act of 1969*, submitted by the department to the U.S. Congress.

3. See W.K.C. Morgan et al., "The Prevalence of Coal Workers' Pneumoconiosis in U.S Coal Miners," *Archives of Environmental Health* 27 (October 1973): 221; W.K.C. Morgan and N. L. Lapp, "Respiratory Disease in Coal Miners," *American Review of Respiratory Disease* 113 (1976): 536. These prevalence data were derived from the periodic X-ray program mandated under the 1969 act.

4. P.L. 91-173, "Federal Coal Mine Health and Safety Act of 1969," sec. 402(b).

5. U.S. General Accounting Office, *Achievements, Administrative Problems, and Costs in Paying Black Lung Benefits to Coal Miners and Their Widows* (Washington, D.C.: U.S. General Accounting Office, 1972), pp. 18, 21.

6. This description of the claims process is based on the author's interviews with claimants and on written evaluations of the program. See U.S. General Accounting Office, *Achievements, Administrative Problems, and Costs*; and "Obtaining Black Lung Benefits in West Virginia and Kentucky Under the Federal Mine Health and Safety Law, A Critique," ARDF Public Interest Report no. 7, photocopy, (Charleston, W. Va.: Appalachian Research and Defense Fund, n.d.).

7. U.S. General Accounting Office, *Achievements, Administrative Problems, and Costs*, p. 2.

8. Ibid., p. 18.

9. These assertions are based on the author's interviews with black lung claimants. See ARDF, "Obtaining Black Lung Benefits."

10. "We Demand Our Rights," mimeographed leaflet in the author's possession.

11. The writings of W.K.C. Morgan and his associates express this point of view. For example, "The presence of severe shortness of breath in a coal miner with simple CWP is virtually always related to nonoccupationally related disease, such as chronic bronchitis or emphysema, rather than to coal mining. . . . Smoking is by far the most important factor in producing respiratory symptoms and a decrease in ventilatory function" (Morgan and Lapp, "Respiratory Disease in Coal Miners," pp. 540–41).

12. G. Jacobsen and W. S. Lainhart, "ILO U / C 1971 International Classification of Radiographs of the Pneumoconioses," *Medical Radiography and Photography* 48, no. 3 (1972): 68.

13. Most physicians would not dispute the assertion that "some miners with simple CWP have extreme functional impairment of the lungs." Disagreement arises over the nature and origin of the disesase process(es) implicated in the impairment: Is it pneumoconiosis, other occupational diseases, or some disease unrelated to work? For a sample of the conflicting viewpoints, see Morgan and Lapp, "Respiratory Disease in Coal Miners"; J. P. Lyons et al., "Pulmonary Disability in Coal Workers' Pneumoconiosis," *British Medical Journal* 1 (March 18, 1972): 713–16; and D. L. Rasmussen and C. W. Nelson, "Respiratory Function in Southern Appalachian Coal Miners," *American Review of Respiratory Disease* 103 (1971): 240–48.

14. The validity of the X-ray as a diagnostic tool for miners' occupational respiratory disease was a central issue in the congressional debate over the 1972 amendments to the Coal Mine Health and Safety Act. See the summary of medical testimony in Senate Committee on Labor and Public Welfare, Subcommittee on Labor, *Legislative History of the Federal Coal Mine Health and Safety Act of 1969 (Public Law 91-173), as Amended through 1974, Including Black Lung Amendments of 1972* (Washington, D.C.: U.S. Government Printing Office, 1975), pt. 2, app., pp. 1773–78.

15. R. Ryder et al., "Emphysema in Coal Workers' Pneumoconiosis," *British Medical Journal* 3 (1970): 481–87; A. Crockett et al., "Post-Mortem Study of Emphysema in Coalworkers and Non-Coalworkers," *Lancet*, 1982, pp. 600–603; "Should Coalworkers Be Compensated for Emphysema?," editorial, *Lancet*, 1983, pp. 626–27; V. Anne Ruckley et al., "Emphysema and Dust Exposure in a Group of

Coal Workers," *American Review of Respiratory Disease* 129, no. 4 (April 1984): 528–32; J. M. Rogan et al., "Role of Dust in the Working Environment in Development of Chronic Bronchitis in British Coal Miners," *British Journal of Industrial Medicine* 30 (1973): 217–26; D. L. Rasmussen, "Impairment of Oxygen Transfer in Dyspneic and Non-Smoking Soft Coal Miners," *Journal of Occupational Medicine* 13 (1971): 300–305; and the discussion in James L. Weeks and Gregory R. Wagner, "Compensation for Occupational Disease with Multiple Causes: The Case of Coal Miners' Respiratory Diseases," *American Journal of Public Health* 76 (January 1986): 58–61.

16. Senate Subcommittee on Labor, *Legislative History . . . as Amended*, pt. 2, app., p. 1777.

17. Ibid., p. 1776.

18. During 1971 and 1972, officials from the Social Security Administration fought the legislative elimination of the X-ray requirement, as contained in the 1972 amendments to the U.S. Coal Mine Health and Safety Act. Ibid., especially pp. 1753–56.

19. See Stanley Joel Reiser, *Medicine and the Reign of Technology* (Cambridge: Cambridge University Press, 1978).

20. See the discussion in Reiser, *Medicine and the Reign of Technology*, pp. 66–68, 189–92. An analysis of this problem as it pertains to diagnosis of CWP may be found in R. B. Reger, H. E. Amandus, and W.K.C. Morgan, "On the Diagnosis of Coalworkers' Pneumoconiosis: Anglo-American Disharmony," *American Review of Respiratory Disease* 108 (1973): 1186–91.

21. Author's interview with disabled coal miner, Logan County, W. Va., September 6, 1978.

22. Author's interview with black lung activist, Raleigh County, W. Va., September 19, 1978.

23. This is a line from the song, "Which Side Are You On?" written by Florence Reece.

24. Author's interview with black lung activist, Raleigh County, W. Va., September 19, 1978.

25. Author's interview with working coal miner, Raleigh County, W. Va., August 24, 1978.

26. Author's interview with black lung activist, Washington, D.C., September 22, 1980.

27. This account is based on local newspaper coverage of the strike. See the Charleston (W. Va.) *Gazette* and *Daily Mail*, the *Raleigh Register* (Raleigh County, W. Va.), and the Beckley (W. Va.) *Post Herald*.

28. Author's interview with black lung activist, Raleigh County, W. Va., September 18, 1978.

29. Ibid.

30. Author's interview with black lung activist, Monongalia County, W. Va., October 11, 1978.

31. Author's interview with black lung activist, Raleigh County, W. Va., September 18, 1978.

32. The *Black Lung Bulletin* was published for about two years, from the summer of 1970 to the summer of 1972. The first four issues highlighted the insurgent struggle within the United Mine Workers; after claims denials began to reactivate the black lung controversy in the fall of 1970, the *Bulletin* focused more attention on the compensation program. *Black Lung Bulletin* 1, no. 1 through *Black Lung Bulletin* 3, no. 3, complete file in the author's possession.

33. Author's interview with black lung activist, Monongalia County, W. Va., October 11, 1978.

34. Author's interview with black lung activist, Raleigh County, W. Va., September 18, 1978.

35. Author's interview with disabled miner, Mingo County, W. Va., September 12, 1978. See also "Earl Stafford: Mountaineer of the Month," *Mountain Call* 1, no. 4 (February 1974): 3–6, 11–12.

36. Author's interview with black lung activist, Raleigh County, W. Va., June 29, 1978.

37. Author's interview with black lung activist and miner's wife, Kanawha County, W. Va., October 4, 1978.

38. Author's interview with disabled miner, Mingo County, W. Va., September 15, 1978.

39. Author's interview with disabled miner, wmingo County, W. Va., September 12, 1978.

40. Ibid.

41. Letter from Orville McCoy to the *Black Lung Bulletin*, July 26, 1971, photocopy in the possession of the author.

42. Author's interview with disabled miner, Raleigh County, W. Va., July 17, 1978.

43. Author's interview with disabled miner, Mingo County, W. Va., September 15, 1978.

44. Author's interview with disabled miner, Logan County, W. Va., September 6, 1978.

45. "We Demand Our Rights," mimeographed leaflet in the author's possession.

46. These demands were put forth in leaflets and in the *Black Lung Bulletin*. See *Black Lung Bulletin* 1, no. 9 (February 1971), p. 3.

47. Senate Subcommittee on Labor, *Legislative History . . . as Amended*, pt. 2, app., 1706–11.

48. *Black Lung Bulletin* 2, no. 1 (June 1971): 4.

49. Author's interview with Appalred staff person, Kanawha County, W. Va., October 4, 1978.

50. Prepared statement of BLA representative, from author's personal files.

51. Author's interview with disabled miner's wife and black lung activist, Raleigh County, W. Va., September 22, 1978.

52. See Senate Subcommittee on Labor, *Legislative History . . . as Amended*, pt. 2, app., pp. 2101–11.

53. This brief analysis of the politicking over the 1972 amendments is based on ibid. and on news reports from the Charleston *Gazette* and the *Washington Post.*

54. See 20 C.F.R., pt. 410; and *Federal Register* 38 (1973): 16962–81, 26042–68.

55. Author's interview with black lung activist, Kanawha County, W. Va., October 1, 1978.

## Chapter 6

1. United Mine Workers of America, *The Year of the Rank and File, Officers' Report to the 46th Constitutional Convention* (Washington, D.C.: United Mine Workers of America, 1973), pp. 7–8.

2. Author's interview with black lung activist, Kanawha County, W. Va., October 8, 1978.

3. U.S. Department of Health, Education, and Welfare, *Black Lung Benefits Program, Third Annual Report on Part B of Title IV of the Federal Coal*

*Mine Health and Safety Act of 1969* (Washington, D.C.: U.S. Government Printing Office, 1974), p. 1.

4. Senate Committee on Labor and Public Welfare, Subcommittee on Labor, *Black Lung Benefits Reform Act, 1976*, Hearings, 94th Cong., 2d sess., 1976, p. 128.

5. *Pichon* v. *Mathews*, 408 F. Supp. 1 (N.D. Ill., February 5, 1976), as quoted in a case summary prepared by Gail Falk, "Additional Federal Court Decisions in Black Lung Cases," UMWA, n.d., photocopy, p. 1.

6. Author's interview with black lung activist, Kanawha County, W. Va., October 8, 1978.

7. See the article by Bill Bishop and Helen Winternitz, "Diagnosis System Stalls Effort to Aid Miners," *Mountain Eagle* (Whitesburg, Ky.), December 18, 1975, pp. 1, 3. The authors reported on a study performed for NIOSH by the Department of Radiology at Johns Hopkins University which concluded that physicians who failed social security's examination for B-readers exhibited "an excessive overreading tendency."

8. U.S. General Accounting Office, *Legislation Authorized Black Lung Benefits without Adequate Evidence of Black Lung or Disability* (Washington, D.C.: U.S. Government Printing Office, 1982), p. 16.

9. *Stewart* v. *Mathews*, CA 75-0146 (W.D. Va., December 19, 1975), as quoted in a case summary prepared by Falk, "Additional Federal Court Decisions in Black Lung Cases," April 1976, p. 3.

10. Ibid.

11. Dr. Donald Rasmussen, whose research and conclusions regarding the disease process of black lung led him to emphasize the diagnostic importance of blood gas tests, was vocal in this criticism. See Donald L. Rasmussen, M.D., "Statement," Sept. 3, 1975, photocopy of transcript in author's possession, p. 15.

12. *Clark* v. *Mathews*, C-2-75-573 (S.D. Ohio, February 17, 1976), case summary prepared by Falk, "Additional Federal Court Decisions in Black Lung Cases," April 1976, p. 3.

13. *Ciaravella* v. *Richardson*, 377 F. Supp. 201 (W.D. Pa., July 19, 1974), case summary prepared by Falk, "Federal Court Decisions in Black Lung Cases," October 1975, p. 4.

14. The July 1, 1973, cutoff applied to claims from living miners. January 1, 1974, was the cutoff date for survivors' claims. See P.L.

92-303, May 19, 1972.

15. U.S. Department of Labor, Employment Standards Administration, *Black Lung Benefits Act of 1972, Annual Report on Administration of the Act during Calendar Year 1976*, Submitted to Congress August 1977 (Washington, D.C.: U.S. Government Printing Office, 1977), p. 21.

16. Ibid., p. 6. On the problems with the federal black lung compensation program during this period, see Senate Subcommittee on Labor, *Black Lung Benefits Reform Act, 1976*, especially pp. 459–517; and U.S. Department Of Labor, OWCP Task Force, *Black Lung Benefits Program*, August 31, 1976.

17. Helen Powell, "In Memory and Honor: Anise Floyd," *Mountain Life and Work*, June 1978, p. 33.

18. Author's interview with black lung activist, Kanawha County, W. Va., October 8, 1978.

19. Author's interview with former UMWA staff person, Washington, D.C., April 11, 1980.

20. Author's interview with former UMWA staff person, Washington, D.C., September 24, 1980.

21. Author's interview with former UMWA staff person, Washington, D.C., April 11, 1980.

22. Author's interview with former UMWA staff person, Washington, D.C., September 22, 1980.

23. Author's interview with former UMWA staff person, Washington, D.C., April 11, 1980.

24. Author's interview with former UMWA staff person, Washington, D.C., September 22, 1980.

25. Author's interview with former UMWA staff person, Washington, D.C., September 24, 1980.

26. For overall analyses of this period in the union's history, see Curtis Seltzer, "The Unions: How Much Can a Good Man Do?" *Washington Monthly*, June 1974, pp. 7–24; Ward Sinclair, "Miners for Democracy: Year One at the UMW," *Ramparts*, June 1974, pp. 37–41, 56–57; Paul John Nyden, "Miners for Democracy: Struggle in the Coal Fields" (Ph.D. diss., Columbia University, 1974), pp. 831–75.

27.The analysis of political divisions within the black lung movement is based on the author's interviews with participants and her direct experiences with the movement during this period.

28. Author's interview with disabled miner, McDowell County, W. Va., August 31, 1978.

29. Author's interview with former UMWA staff person, Kanawha County, W. Va., October 1, 1978.

30. Author's interview with black lung activisit, Fayette County, W. Va., July 26, 1978.

31. Author's interview with former UMWA staff person, Kanawha County, W. Va., October 1, 1978.

32. Ibid.

33. Ibid.

34. Author's interview with disabled miner, McDowell County, W. Va., August 7, 1974.

35. Author's interview with working miner, Raleigh County, W. Va., August 24, 1978.

36. "Resolutions Passed at Pikeville, Kentucky, November 17–18, by Members of Black Lung Associations in Tennessee, Kentucky, West Virginia, and Virginia," photocopy in the author's personal files.

37. "Fight for Black Lung Benefits," leaflet in the author's personal files.

38. U.S. Department of Labor, Bureau of Labor Statistics, *Collective Bargaining in the Bituminous Coal Industry*, BLS Report no. 514 (Washington, D.C.: U.S. Government Printing Office, 1977).

39. U.S. Department of Labor, Employment Standards Administration, *Black Lung Benefits Act, Annual Report on Administration of the Act during Calendar Year 1979*, Submitted to Congress 1980 (Washington, D.C.: U.S. Government Printing Office, 1980), p. 21.

40. *Business Week*, November 16, 1974, p. 33.

41. See U.S. Department of Labor, BLS, *Collective Bargaining in the Bituminous Coal Industry*, p. 5.

42. These are the words of Joseph Brennan, BCOA president, as quoted in the Charleston *Gazette*, July 27, 1976, p. 1.

43. Barbara Ellen Smith, "Coal Miners and the Federal Courts: Wildcat Strikes in West Virginia, 1972–1977" (paper, Brandeis University, 1978). A survey of federal district court records in southern West Virginia revealed a marked dropoff in coal companies' use of the courts after the 1976 wildcat strike.

44. Author's interview with former UMWA staff person, Washington, D.C., September 24, 1980.

## Chapter 7

1. As of 1984, the basic monthly black lung benefit for a single claimant was $315.60. For miners or survivors with one dependent, the rate was $473.30. Statistics on black lung claims are from the U.S. Department of Labor, Employment Standards Administration, and the U.S. Department of Health and Human Services, Social Security Administration. See U.S. Department of Labor, Employment Standards Administration, *Black Lung Benefits Act, Annual Report on Administration of the Act during Calendar Year 1982* (Washington, D.C.: U.S. Government Printing Office, 1984).

2. The classic position in the sociology of knowledge was articulated by Karl Mannheim. See his *Ideology and Utopia* (New York: Harcourt, Brace and World, 1936) and *Essays on the Sociology of Knowledge* (London: Routledge and Kegan Paul, 1952). See the summary of this literature in Michael Mulkay, *Science and the Sociology of Knowledge* (London: George Allen and Unwin, 1979).

3. Mulkay, *Science*, p. 112.

4. There has been a recent explosion of psychological, philosophical, ethnomethodological, and other literature in this vein. See the summary in Mulkay, *Science*; Karin D. Knorr-Cetina and Michael Mulkay, *Science Observed* (London: Sage Publications, 1983); Hilary Rose and Steven Rose, *The Political Economy of Science* (London: Macmillan Press, 1976); idem, *The Radicalisation of Science: Ideology of / in the Natural Sciences* (London:Macmillan Press, 1976); Bob Young, "Science *Is* Social Relations," *Radical Science Journal*, no. 5 (1977), pp. 65–129; Evelyn Fox Keller, *Reflections on Gender and Science* (New Haven, Conn.: Yale University Press, 1985).

5. Karin D. Knorr-Cetina, *The Manufacture of Knowledge: An Essay on the Constructivist and Contextual Nature of Science* (Oxford: Pergamon Press, 1981); and the articles by Bruno Latour and Karin D. Knorr-Cetina in Knorr-Cetina and Mulkay, *Science Observed*.

6. Keller, *Reflections on Gender and Science*; Sandra Harding, *The Science Question in Feminism* (Ithaca: Cornell University Press, 1986).

7. See Evelyn Fox Keller, "The Force of the Pacemaker Concept in Theories of Aggregation in Cellular Slime Mold," in Keller, *Reflections on Gender and Science*; R. C. Lewontin, Steven Rose, and Leon J. Kamin, *Not in Our Genes: Biology, Ideology, and Human Nature* (New York: Pantheon Books, 1984). A related but quite different debate occurs within theoretical physics, which, in the very principles of quantum mechanics, calls into question the assumption that nature is objectifiable and knowable. This has generated an intriguing body of popular literature that explores the philosophical similarities between theoretical physics and Eastern religion. See Fritjof Capra, *The Tao of Physics* (New York: Bantam Books, 1977); and Gary Zukav, *The Dancing Wu Li Masters: An Overview of the New Physics* (New York: Bantam Books, 1980).

8. See the summary in Lewontin, Rose, and Kamin, *Not in Our Genes*, chap. 3, pp. 37–61; and Rose and Rose, *Political Economy of*

9. See the discussions of Darwin in Mulkay, *Science*, pp. 100–109; and Lewontin, Rose, and Kamin, *Not in Our Genes*, pp. 48–51.

10. See Lewontin, Rose, and Kamin, *Not in Our Genes*, especially chap. 3, pp. 37–61. The "social constructionist" perspective on science does not necessarily imply that all forms of knowledge, from witchcraft to racism to astrophysics, are equally valid, though some analysts arrive at this conclusion. The problem with this extreme view is that it obliterates the material world as a determinant of the range of scientific interpretation. It also ignores the special character of science as the form of knowledge that systematically invites material reality to construct and constrain it.

11. *Steadman's Medical Dictionary* (Baltimore: Williams and Wilkins, 1976), p. 401.

12. See Lewontin, Rose, and Kamin, *Not in Our Genes*; Lesley Doyal with Imogen Pennell, *The Political Economy of Health* (London: Pluto Press, 1979); and Meredeth Turshen, "The Political Ecology of Disease," *Review of Radical Political Economics* 9, no. 1 (1977): 45–60.

13. See Doyal with Pennell, *Political Economy of Health*, pp. 29–30.

14. Howard S. Berliner and J. Warren Salmon, "The Holistic Health Movement and Scientific Medicine: The Naked and the Dead," *Socialist*

*Review* 9, no. 1 (January–February 1979): 31–52.

15. Ibid., and Susan Reverby, "A Perspective on the Root Causes of Illness," *American Journal of Public Health* 62, no. 8 (August 1972): 1140–42.

16. See Talcott Parsons, *The Social System* (New York: Free Press, 1951). Parsons's conceptualization of the sick role was neither class nor historically specific.

17. It would seem from this conclusion that the role and nature of scientific / technical knowledge might be a fruitful area of exploration for theorists concerned about the "relative autonomy" of the state.

18. Many of the physicians who argued for a restrictive definition of disabling CWP may well have been politically conservative. This could partly explain their affinity for a restrictive perspective on disease, but it does not explain the epistemological origins of that perspective.

19. See Lewontin, Rose, and Kamin, *Not in Our Genes.*

20. This paraphrases the questions posed by Michel Foucault throughout his exploration of the social construction and regulation of the body. See his *The Birth of the Clinic: An Archaeology of Medical Perception* (New York: Vintage Books, 1975) and *Discipline and Punish: The Birth of the Prison* (New York: Vintage Books, 1979).

21. Statistics on dust samples, maintained by the U.S. Mine Health and Safety Administration, indicate that respirable dust underground has been reduced to an extremely low level. (A summary printout of dust samples taken from October 1, 1984, through September 30, 1985, by the coal operators reveals average dust concentrations below 1 $mg/m^3$ for most underground occupations. Data were supplied by the U.S. Mine Health and Safety Administration.) Many miners dispute the accuracy of these samples and the associated conclusion that dust is being adequately controlled. See "Why Aren't Companies Controlling Dust?" UMW *Journal*, February 1984, pp. 3–9.

22. U.S. General Accounting Office, *Legislation Allows Black Lung Benefits to Be Awarded without Adequate Evidence of Disability* (Washington, D.C.: U.S. Government Printing Office, 1980), p. 8.

23. W.K.C. Morgan, "Respiratory Disability in Coal Miners," *Journal of the American Medical Association 243*, no. 23 (June 20, 1980): 2403.

24. William R. Barclay, "Black Lung Benefits," editorial, *Journal of the American Medical Association* 243, no. 23 (June 20, 1980): 2427.

25. U.S. General Accounting Office, *Legislation Authorized Black Lung Benefits without Adequate Evidence of Black Lung or Disability* (Washington, D.C.: U.S. Government Printing Office, 1982), p. iii.

26. U.S. Department of Labor, Employment Standards Administration, *Black Lung Benefits Act, Annual Report on Administration of the Act during Calendar Year 1983* (Washington, D.C.: U.S. Government Printing Office, 1986), p. 23.

# Index